COMMON FLORA

OF THE

PLAYA LAKES

COMMON FLORA
OF THE
PLAYA LAKES

DAVID A. HAUKOS
AND LOREN M. SMITH

PHOTOGRAPHS BY RUSSELL T. WIPFF

TEXAS TECH UNIVERSITY PRESS

This book was set in Zapf Humanist Demi and Bank Gothic and printed on acid-free paper that meets the guidelines for permanence and durability of the Committee on Production Guidelines for Book longevity of the Council on Library Resources. ∞

Photographs appearing on pages ii-iii, 5, and 140 (bottom) by Wyman Meinzer, courtesy U.S. Fish and Wildlife Service; pages 56 (top) and 154 (top), photographs by John Taylor; page 48, photograph by Lynn A. Nymeyer; and page 116, photograph by Kelly Kindscher.

Design by Rob Neatherlin

Printed and manufactured in Hong Kong by Sing Cheong Printing Co. Ltd.

Library of Congress Cataloging-in-Publication Data
Haukos, David A.
      Common flora of the playa lakes / David A. Haukos and Loren M. Smith; photographs by Russell T. Wipff.
          p. cm.
      Includes index.
      ISBN 0-89672-388-7 (pbk. : alk. paper)
      1. Wetland plants—Texas—Identification. 2. Wetland plants—Southwestern States—Identification. 3. Wetland plants—Texas—Pictorial works. 4. Wetland plants—Southwestern States—Pictorial works. I. Smith, Loren M. II. Title.
QK188.H395 1997
581.7'4'097648—dc21   97-13881
                                                                         CIP

97 98 99 00 01 02 03 04 05/ 9 8 7 6 5 4 3 2 1

Texas Tech University Press
Box 41037
Lubbock, Texas 79409-1037 USA
800-832-4042
ttup@ttu.edu

CONTENTS

ACKNOWLEDGMENTS

This book would not have been possible without the efforts of many agencies and people. Funding for this project was provided by the Division of Refuges and Wildlife, Region 2 of the U.S. Fish and Wildlife Service, the Department of Range, Wildlife, and Fisheries Management, Texas Tech University, and the Playa Lakes Joint Venture (especially the states of Texas, New Mexico, Oklahoma, Kansas, and Colorado, and Phillips Petroleum Company). Leslie Dierauf, Jeff Haskins, and John Cornely coordinated project funding.

Russell Wipff, Michael Whitson, and Jennifer Warren assisted in the collection of field data, preparation of voucher specimens, and a variety of other activities. Russell Pettit, Jay Wipff, and Stanley Jones verified species identifications and identified unknown specimens. Voucher specimens are stored in the herbaria of the Department of Range, Wildlife, and Fisheries Management, Texas Tech University and the Botanical Research Center, Bryan, Texas. Kay Arellano typed much of the original manuscript. Arthur Elliot and Charles Werth provided access to the herbarium in the Biology Department of Texas Tech University. I. M. Ortega constructed the map of the Playa Lakes Region.

The Natural Resources Conservation Service assisted in locating some owners of study playas. We thank the multitude of private landowners that allowed us access to their playas. Forest Service personnel on Comanche, Cimarron, Kiowa, and Rita Blanca National Grasslands provided playa locations, history, and access. Lynn Nymeyer, manager of the Buffalo Lake National Wildlife Refuge, provided access to refuge wetlands and some photographs that are included in this book.

Kelly Kindscher, Kansas Biological Survey and Shaun Vickers, Natural Resources Conservation Service, provided unpublished species lists and other information on playas in unsurveyed areas of Kansas. Susan Schott, University Press of Kansas, provided guidance and permission to include portions of the glossary and species descriptions from the *Flora of the Great Plains* (Great Plains Flora Association 1991). We are indebted to Gary Larsen, South Dakota State University, Warren Hagenbuck, U.S. Fish and Wildlife Service, and Gary Valentine, Natural Resources Conservation Service for their constructive, insightful, and thought-provoking reviews of the manuscript. Porter B. Reed, Jr., National Wetland Inventory, provided the current unpublished hydrophytic category list for plants in the Playa Lakes Region. However, we take full responsibility for any errors contained herein.

Jane Jacobi graciously copyedited the manuscript. We would especially like to thank Carole Young and Texas Tech University Press for their hard work preparing the manuscript for publication.

INTRODUCTION

Twenty-five to thirty thousand playa wetlands occur in southeastern Colorado, southwestern Kansas, western Oklahoma, eastern New Mexico, and northwest Texas (see map) (Guthery et al. 1981, Osterkamp and Wood 1987). Knowledge of the composition and dynamics of these playa plant communities has been limited to a few local studies. Comprehensive descriptions of wetland flora are needed by land and wildlife managers to make sound resource decisions. This information is also needed by state and federal agencies for regulatory purposes. Although this type of information exists for most other major wetland ecosystems, it did not exist, until this report, for playa wetlands in the Southern Great Plains.

The primary purpose of this book is to provide wildlife biologists, land managers, regulatory agencies, and other individuals simply interested in wetland plants with a quick, accurate, photographic guide to common plant species found in playa wetlands. Our first objective is to provide the information and photos necessary for easy identification of common playa species in the field, rather than going through a "keying out" process. However, if any question arises as to species determination, we recommend the detailed key in *Flora of the Great Plains* (Great Plains Flora Association 1991). A second objective is to provide a comprehensive list of plant species found in playas from this and other studies for biologists and regulatory agencies. Finally, because the watersheds and basins of most playas have been altered extensively since 1920 (Luo et al. 1997), changing hydrologic conditions, we wanted to provide a baseline dataset that researchers can use to evaluate future changes in playa floral communities. These objectives were accomplished largely through an extensive floristic survey.

We attempted to determine plant species occurrence and distribution in playas by conducting a regional survey of plant communities and consolidating available data on species found in playas. We conducted a regional floristic survey of playa wetlands during 1995. We randomly surveyed 1% of the playas in 40 counties containing more than 100

The larger number associated with each county in Colorado, Kansas, Oklahoma, New Mexico, and Texas is the number of playas in the county according to Guthery et al. (1981). The smaller number associated with each county represents the number of playas we surveyed in the 1995 study.

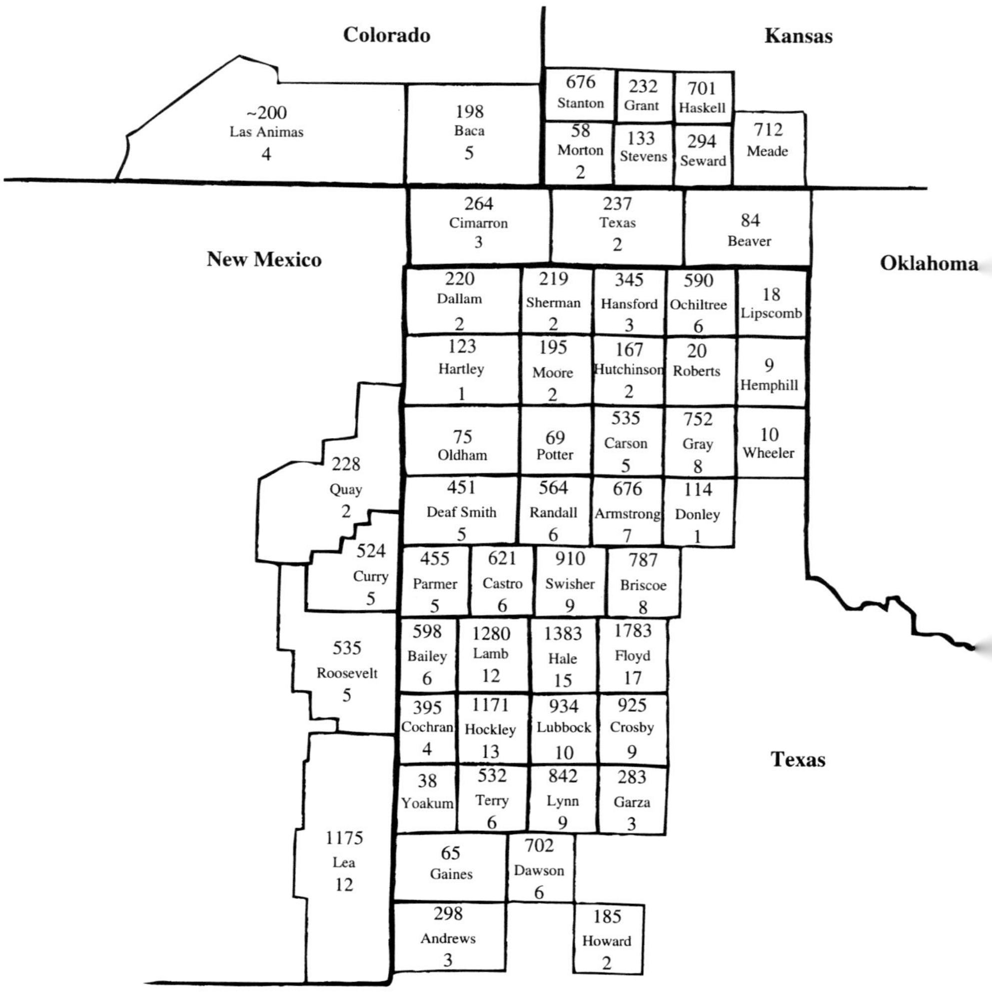

Colorado

Kansas

New Mexico

Oklahoma

Texas

County	Large #	Small #
Las Animas	~200	4
Baca	198	5
Stanton	676	
Grant	232	
Haskell	701	
Morton	58	
Stevens	133	
Seward	294	
Meade	712	
Cimarron	264	3
Texas	237	2
Beaver	84	
Dallam	220	2
Sherman	219	2
Hansford	345	3
Ochiltree	590	6
Lipscomb	18	
Hartley	123	1
Moore	195	2
Hutchinson	167	2
Roberts	20	
Hemphill	9	
Oldham	75	
Potter	69	
Carson	535	5
Gray	752	8
Wheeler	10	
Deaf Smith	451	5
Randall	564	6
Armstrong	676	7
Donley	114	1
Quay	228	2
Curry	524	5
Parmer	455	5
Castro	621	6
Swisher	910	9
Briscoe	787	8
Bailey	598	6
Lamb	1280	12
Hale	1383	15
Floyd	1783	17
Roosevelt	535	5
Cochran	395	4
Hockley	1171	13
Lubbock	934	10
Crosby	925	9
Yoakum	38	
Terry	532	6
Lynn	842	9
Garza	283	3
Lea	1175	12
Gaines	65	
Dawson	702	6
Andrews	298	3
Howard	185	2

playas (Guthery et al. 1981) in Texas, New Mexico, Oklahoma, and Colorado (Table 1, map). Two playas were surveyed in Kansas. Attempts to survey other playas in Kansas were unsuccessful because we were unable to: (1) locate playas that had not been filled or cultivated and (2) obtain landowner permission for access to located playas. Information from Kindscher (1994) was used to complement our surveys in Kansas.

We surveyed 233 playas with step-point samples along two transects in each playa. The first transect started from the southeast corner of the playa bisecting the playa at a 45° angle to the west boundary, and then the second transect traversed the playa floor at a 45° angle, to the northeast corner. The playa boundary was determined based on changes in soil color and topography (Luo et al. 1997). All but 11 playas were surveyed twice during the growing season: late spring-early summer (15 May–30 June) and mid-late summer (15 July–31 August) to account for cool- and warm-season species. Cultivation between survey 1 and survey 2 accounted for those playas that were only surveyed once. We started each survey period in the southern portion of the region, working northward, and completed each survey period in Kansas and Colorado.

All known species were identified in the field with collection of voucher specimens for verification. Unknown species were collected for future identification.

The soil moisture condition of each playa was categorized prior to each survey as flooded (standing water over >50% of playa floor), moist (standing water over <50% of playa floor or sufficient topsoil moisture to form and maintain a soil ball), or dry (insufficient topsoil moisture to form a soil ball). Dominant (>50%) surrounding land use (cropland, rangeland) was determined for each sampled playa. Uplands with Conservation Reserve Program (CRP) land were classified as cropland. The CRP is a U.S. Department of Agriculture program whereby highly erodible cropland was retired for 10 years by planting it with permanent cover. In the Playa Lakes Region, cover was mainly perennial grasses. Because most land in this region was enrolled by 1988 (Berthelsen et al. 1989) and was previously exposed to intensive cultivation and associated sedimentation (Luo et al. 1997), we continued to classify it as cropland.

A total of 172,589 step-point samples was taken in surveyed playas; 61.2% in dry playas, 10.4% in moist playas, and 28.3% in flooded playas. We completed 459 playa surveys (224 playas twice and 11 playas once):

Table 1. Playas surveyed for plant community composition in the Playa Lakes Region, 1995.

State/County	Number surveyed	Average size ha (acres)		Number surrounded by rangeland	Number surrounded by cropland
COLORADO					
Baca	5	9.8	(24.2)	5	–
Las Animas	4	21.8	(53.7)	4	–
KANSAS					
Morton	2	5.2	(12.8)	2	–
OKLAHOMA					
Cimmaron	3	56.1	(138.6)	3	–
Texas	2	20.7	(51.0)	1	1
NEW MEXICO					
Curry	5	16.0	(39.5)	4	1
Lea	12	2.9	(7.3)	12	–
Roosevelt	5	8.2	(20.3)	5	–
Quay	2	13.8	(34.1)	2	–
TEXAS					
Andrews	3	8.3	(20.5)	3	–
Armstrong	7	15.7	(38.8)	3	4
Bailey	6	6.7	(16.5)	2	4
Briscoe	8	8.0	(19.8)	8	–
Carson	5	45.5	(112.3)	2	3
Castro	6	22.3	(55.1)	–	6
Cochran	4	5.3	(13.0)	2	2
Crosby	9	20.1	(49.7)	2	7
Dallam	2	3.2	(7.8)	2	–
Dawson	7	6.1	(15.1)	1	6
Deaf Smith	5	14.1	(34.8)	1	4
Donley	1	17.0	(42.0)	1	–
Floyd	18	14.5	(35.9)	4	14
Garza	3	16.3	(40.3)	2	1
Gray	8	36.3	(89.7)	6	2
Hale	15	15.5	(38.4)	3	12
Hansford	3	52.9	(130.7)	–	3
Hartley	1	43.7	(107.9)	1	–
Hockley	13	5.4	(13.4)	6	7
Howard	2	6.6	(16.3)	–	2
Hutchinson	2	14.9	(36.7)	–	2
Lamb	12	13.0	(32.1)	–	12
Lubbock	10	8.1	(20.0)	1	9
Lynn	9	6.8	(16.8)	2	7
Moore	2	22.7	(55.9)	2	–
Ochiltree	6	19.7	(48.5)	–	6
Parmer	5	18.2	(44.8)	–	5
Randall	6	9.3	(23.0)	4	2
Sherman	2	42.2	(104.2)	–	2
Swisher	9	12.3	(30.4)	3	6
Terry	6	5.3	(13.1)	2	4
Total/Average	235	14.7	(36.4)	101	134

There are 25,000–30,000 playas on the Southern Great Plains. These freshwater wetlands greatly increase the biodiversity of the region.

305, 49, and 105 surveys in dry, moist, and flooded conditions, respectively. Cropland was the surrounding land use of 134 playas; the remaining 101 were surrounded by rangeland.

In 1995, most of the region we surveyed was in the third consecutive year of below-average precipitation. This probably skewed the results of our survey by increasing the prevalence of more xeric-type plants than would have occurred in years with precipitation approaching average (but average years are rare in the Southern Great Plains!).

Photos and detailed descriptions of plant species that occurred in greater than 5% of the 235 surveyed playas are included in this guide. We also included selected species that were representative of certain hydrological conditions in playas, despite their occurrence in fewer than 5% of surveyed playas. A complete list of all species found in playas during this and earlier studies is provided in the appendix.

HISTORY AND ECOLOGY OF PLAYA LAKES

Playa lakes are unique, freshwater wetlands of the High Plains region of the Southern Great Plains. The Playa Lakes Region (PLR) includes approximately 36.2 million ha (89.5 million acres; 140,000 square miles) of southwestern Kansas, southeastern Colorado, the panhandle of Oklahoma, eastern New Mexico, and the panhandle and Southern High Plains (SHP) of Texas. Estimates of the number of playas in the PLR range from 25,390 to 37,000 (Reddell 1965, Guthery and Bryant 1982, Osterkamp and Wood 1987). The SHP or Llano Estacado (that area of the PLR south of the Canadian River) of Texas and New Mexico contains most of the playas (ca. 85%) with a playa density of nearly 1 per 2.6 km^2 (1 per square mile) (Guthery et al. 1981).

Although playas provide greater than 160,000 ha of wetland habitat, they occupy only approximately 2% of the total landscape (Haukos and Smith 1994a). The Southern Great Plains was originally short- and mid-grass prairie. Domestic cattle began to graze the region in the 1870s and most crops started to be cultivated during the 1920s. Development of irrigation technology during World War II enabled the current use of the Ogallala Aquifer, leading to expanded cultivation in the PLR since the late 1940s. The PLR is one of the most agriculturally impacted regions in the Western Hemisphere. Cotton, wheat, corn, grain sorghum, and various vegetables are dominant crops, and livestock grazing is practiced on the uncultivated areas (Bolen et al. 1989). Since 1986, area farmers have been encouraged by the United States Department of Agriculture to replace crops with permanent perennial cover on highly erodible land in exchange for annual rental payments to reduce soil erosion—the Conservation Reserve Program (CRP). The PLR had the highest density of CRP land in the nation (Berthelsen et al. 1989).

The region is influenced climatically by the Chihuahuan Desert to the southwest and more mesic prairies to the north and east. The ancient cool and wet climate of the area changed during the middle Holocene

Playas are the major topographical relief for much of the High Plains portion of the Southern Great Plains.

(6000–4500 BP) to warm and dry (Holliday 1991). Currently, the climate is subhumid continental, with average annual precipitation ranging from 35 cm (14 inches) in the west to 63 cm (25 inches) in the east. Precipitation occurs primarily as heavy, localized thunderstorms during May through September (Bolen et al. 1989). Drought is a natural and common occurrence (Holliday 1991).

Temperature fluctuates widely and frequently, with recorded temperatures from below –4° C to above 50° C (–20° F–120° F). Average annual potential evaporation can exceed 250 cm (100 inches) per year, especially in southern areas (Bolen et al. 1989). The growing season averages 140 days in the north to 220 days in south.

Playas are the most significant topographical feature and surface hydrological attribute in the Southern Great Plains (Osterkamp and Wood 1987). Playa watersheds are closed systems, with playa floors representing the deepest point of the watershed (Bolen et al. 1989). Unlike most wetlands, playas have little elevational change as one proceeds across them in a horizontal gradient; playa floors are flat (Luo et al. 1997). Playa floors are defined by the presence of a hydric, vertisol clay soil, usually Randall clay (Allen, et al. 1972). Playas average 6.3 ha (15.5 acres) in surface area (Guthery and Bryant 1982), with 87% of all playas being less than 12 ha (30 acres). Average size of playas increases from southwest to northeast, similar to precipitation patterns (Grubb and Parks 1968, Allen et al. 1972).

Weather patterns of the region control the hydrology and formation of playas. The localized nature of precipitation and prolonged dry periods result in dramatic and unpredictable natural wet-dry cycles in playas. Few playas are directly connected to groundwater, and historically, playas were totally dependent upon precipitation runoff to flood (Bolen et al. 1989). The advent of crop irrigation and the resultant runoff (tailwater) have provided another source for flooding of playas. Indeed, in some cultivated areas of the PLR, playas have become more frequently flooded than in the past because of increased precipitation runoff from cultivated areas and an influx of tailwater from irrigated croplands.

The various theories on playa formation are controversial. Historically, relatively simple processes ranging from deflation, animal activity, dissolution of underlying evaporites, and leaching of calcic soils were used to explain playa formation (Osterkamp and Wood 1987, Gustavson et al. 1995). However, playa formation and maintenance is a dynamic and continuing process of interacting geomorphic, pedogenic, and hydrochemical mechanisms (Osterkamp and Wood 1987, Gustavson et al. 1995).

There are currently two prominent theories attempting to explain the formation of playas. Osterkamp and Wood (1987) and Wood and Osterkamp (1987) proposed that dissolution of calcic soils and calcretes that underlie the High Plains is the principal cause of playa formation. They developed a model of formation based on hydrologic and geomorphic processes that were initiated when water collected in depressions on the prairie. According to this model, when water collected in depressions, it percolated downward, carrying organic matter. Oxidation of organic matter formed carbonic acid, which dissolved the caliche layer. Dissolution increased permeability of surface waters, allowing increased transport of solutes, particulate rock, and organic matter. This increase in material transport, combined with surrounding land subsidence (due to dissolution of caliche) or compaction of remaining caliche and clastic beds, deepened and expanded basins in a circular fashion from a central point. As basins increased in size, the amount of clay-sized material entering the basin increased; some of this material could not be removed by downward transport, resulting in development of a nearly impermeable basin floor.

Plant communities may frequently change during a growing season in response to the variable playa environment.

Gustavson et al. (1995) proposed that no single process was responsible for development of playa basins. They postulated that a combination of depositional, pedogenic, geomorphic, and hydrologic processes contributed to the formation of playas. They believed small ephemeral ponds developed on the landscape from deflation, subsidence over salt dissolution, or differential compaction. Over time, centripetal drainage carried runoff and sediment into the depression. Periodic elimination of stabilizing vegetation by flooding events may have resulted in wind deflation of sediments following drying periods. Eolian sediment accumulated around the depression, and calcic soils developed in the interdepressional areas. Centripetal drainage enlarged the playa basin by slope erosion. Aquifer recharge through the playa minimized the accumulation of calcium carbonate, which accumulated elsewhere on the High Plains. The depression continued to enlarge and accumulate sediment due to centripetal erosion. During dry periods, sediment in the depression was constantly subject to deflation. The result was that upland areas grew vertically at rates faster than that of playa floors (Gustavson et al. 1995). The playa basins remained stable landforms, because playa sediments are subject to wind deflation and few carbonates accumulate within playa sediments.

Despite being surrounded by intensive agricultural activities, playas continue to perform many functions beneficial to humans and biota of the region. Playas collect flood waters for approximately 90% of the region, and are particularly important for catchment of storm-water runoff in urban areas (Hertel and Smith 1994). Playas are critical recharge points to the southern portions of the Ogallala Aquifer, filtering and recharging 20–80% of collected water to the aquifer (Osterkamp and Wood 1987, Wood and Osterkamp 1987, Zartman 1987, Zartman et al. 1996); thus, they are critical for the continued maintenance of the aquifer. Many farmers use playas as a part of their irrigation systems, by recycling runoff irrigation water and pumping collected precipitation runoff back onto surrounding crops. Greater than 70% of playas larger than 4 ha (10 acres) have been modified by construction of pits to concentrate water for recycling of runoff water for irrigation purposes (Guthery and Bryant 1982). Many playas are used for catchment and storage of industrial, municipal, and animal feedlot wastewater runoff. Playas provide livestock forage, especially in periods of prolonged droughts.

Because of intensive agriculture in areas surrounding playas, these wetlands are the principal remaining native habitat for wildlife in the region. Many landowners lease their playas for hunting, providing an economic return for the management of wildlife. Most, if not all, species of wildlife in the region use playas, and many species are dependent on playas for their existence. Indeed, the biodiversity of the PLR is dependent upon playas (Haukos and Smith 1994a). Nearly 200 species of birds have been identified in playa wetlands. Nine species of amphibians, which consume a multitude of agricultural pest insects, would not exist in much of the region without playas (Anderson 1997). A minimum of 37 species of mammals have been associated with playas. Several species of reptiles use playas throughout the year.

The altered hydroperiod resulting from cultivation and the unstable, unpredictable weather in the PLR has influenced vegetation occurrence and distribution in playas. Usually, there exists a short period of time each year during which suitable environmental conditions exist for most plant species to germinate, grow, and reproduce. Therefore, most plant species now occurring in playas are annuals or short-lived perennials that rapidly respond to appropriate soil temperature and moisture conditions to become established in playas. Additionally, soil nutrients limiting plant growth vary depending on soil moisture, with nitrogen and phosphorus potentially limiting during wet and dry years, respectively (Haukos and Smith 1996).

The ephemeral nature of water in playas may enhance floristic diversity, which in turn leads to increased faunal diversity (Haukos and Smith 1994a). Conversely, however, these same ephemeral patterns rarely result in a stable flora in playas throughout a growing season. When flooded, playas contain submergent and emergent aquatic species representative of other United States freshwater wetlands (Haukos and Smith 1993). Playas with moist-soil conditions (saturated soil with no standing water) develop communities dominated by annuals capable of producing large quantities of seed (Haukos and Smith 1993). Dry playas are characterized by plant species more commonly found in surrounding uplands, including species of the native prairie. As the environment changes throughout the growing season, so does community structure (Haukos and Smith 1993).

Conversion of the Southern Great Plains from native prairie to agricultural uses has affected the hydrology and vegetation in playas. Reed (1930) described 25 plant species in playas and noted that playa vegetation differed from that of the surrounding upland. Rowell (1971, 1981) described 69 plant species in playas. Haukos and Smith (1993) added 17 more species, many of which would have been expected to occur in the native prairie surrounding the playas. Combining these published reports with plant surveys by Hoagland (1991) in Colorado, Curtis and Beierman's (1980) regional survey, Cushing et al. (1993) and Johnston (1993) on the Department of Energy's Pantex Reservation near Amarillo, and Kindscher and Lauver (1993), Kindscher (1994), and the Natural Resources Conservation Service in Kansas leads to a total of 282 species previously reported from playas (Appendix).

Our 1995 survey provides a regional composition of plant communities in playas. We identified 195 species, adding 64 species to the existing reported 282 species—a total of 346 plant species now reported in playa wetlands. A word of caution, however: several species identified in playas by Curtis and Beierman (1980) have not been confirmed in the Southern Great Plains according to Great Plains Flora Association (1991). Several of the species additions from our study also may represent exotic species originally planted into CRP land.

Moist-soil vegetation (annuals that germinate in saturated soil) provides high quality wildlife habitat.

Because playas are closed basins, little interplaya transport of seed exists within a season (Bolen et al. 1989). As a result, seasonal development of vegetation communities is primarily from underlying seed banks (viable seed in and on the soil) with regeneration by tubers and rhizomes by some species. The composition of seed banks is similar (i.e., little zonation) throughout a playa basin (Haukos and Smith 1994a), enabling the establishment of a plant species anywhere in the wetland given appropriate environmental conditions for germination and growth. Therefore, vegetation present in a naturally functioning playa at any point is directly related to the moisture regimes of previous years, which create the seed bank, and the moisture regime of the current growing season, which regulates germination and seedling growth (Haukos and Smith 1993). In addition to maintaining seed for vegetation, playa seed banks also serve as a repository for species of the original native prairie.

Playa vegetation communities are dependent upon disturbance (wet-dry cycles) to rejuvenate themselves and maximize biological production. Without natural wet-dry cycles, diversity of vegetation would be reduced. Reduction of diversity of playa vegetation would directly result in elimination or reduction of many wildlife species dependent on playas. Therefore, it is essential that playas be allowed to function naturally within the agricultural landscape, and conservation efforts strive for protection of the playa hydroperiod to maintain the diversity of native vegetation.

Sedimentation from cropland has been identified as the major threat to the integrity of playa ecosystems (Luo et al. 1997). Luo et al. found that playas with cultivated watersheds contained more sediment than those surrounded by range, and had lost much of their original volume. If similar sedimentation rates continue, most playas will lose their original volume within 100 years (Luo et al. 1997). Such accumulation of sediment affects playa hydrology by decreasing the hydroperiod and resulting in increasingly more xeric conditions.

Because most playas are surrounded by croplands, they can potentially receive runoff containing herbicides and pesticides. However, only a few studies have examined the impact of these chemicals on playa wildlife (Wallace 1984, Flickinger and Krynitsky 1987). Price et al. (1989) found that tebuthiuron (a herbicide used on rangelands) would adversely impact playa algal communities when applied at rates greater than 2.2 kg/ha. Residues of triazine herbicides and aldicarb insecticides also have been found in playas (Mollhagen et al. 1993).

Dumping of contaminated water from oil field operations has caused widespread mortality in avian populations (Nelson et al. 1983). However, the impacts on plants, invertebrates, and other wildlife species are largely unknown. Although losses were once thought to occur mainly in a few saline lakes, the problem is now known to be more widespread in the PLR (Nelson et al. 1983).

Modification of playas by construction of irrigation pits to concentrate irrigation and precipitation runoff may allow surface water to remain during periods when playas would normally be dry (Bolen et al. 1989). However, the ultimate impacts of concentrating water, which otherwise would be spread out over a shallow littoral basin, on establishment of plant species is unknown. It certainly decreases hydroperiod of the entire basin that, in turn, changes the composition of the vegetation by allowing the establishment of perennial wetland vegetation (e.g., cattails, *Typha* spp. and bulrushes, *Scirpus* spp.) in the pit area.

Overgrazing of playas and their watersheds by cattle can be detrimental to biodiversity in playa ecosystems (Haukos and Smith 1994a). Guthery et al. (1982) found that grazing in playa basins reduced plant

Vegetation (and associated water) in and around playas forms the dominant type of wildlife habitat for much of the Playa Lakes Region.

diversity and increased species such as buffalo grass (*Buchloe dactyloides*), cocklebur (*Xanthium strumarium*), and bur ragweed (*Ambrosia grayi*). Guthery et al. (1980) suggested elimination of grazing around playas to increase and improve nesting habitat for ring-necked pheasant (*Phasianus colchicus*) and waterfowl. Overgrazing a playa watershed eliminates desirable cover for most species of wildlife and increases erosion, which will increase turbidity and sedimentation.

There are several keys to the conservation of playas. First and foremost, the importance of playas to local landscape heterogeneity and regional and continental biodiversity must be recognized and understood by lawmakers, governmental agencies, conservation groups, agricultural organizations, and local citizens. Because greater than 99% of playas are privately owned, most management-oriented research and conservation efforts (other than acquisition) dealing with playas should be acceptable to private landowners and compatible with local agricultural activities. When private landowners understand the beneficial functions of playas, efforts to conserve playas are enhanced. Therefore, education programs are crucial to playa conservation. Perhaps the best approach to active conservation of playas by landowners is through watershed scale protection of natural events that influence and drive playa functions and through reduction of anthropogenic threats to the playa.

The U.S. Fish and Wildlife Service and the Natural Resources Conservation Service as well as the five state wildlife conservation agencies, have programs designed to assist private landowners in the management of playas. Vehicles for conservation of playas in 1997 include the Wetland Reserve Program (WRP) and Wildlife Habitat Incentive Program (WHIP) of the Natural Resouces Conservation Service, the Partners for Wildlife Program (PWP) and Challenge Cost-Share Program of the U.S. Fish and Wildlife Service, and the private-land program of the conservation agencies of each state. Many of these programs are designed to assist private landowners in the conservation and management of playas; however, they have only recently been activated in the area and have made limited progress. Because of the importance of playas to continental wetland wildlife, the Playa Lakes Joint Venture (a conglomeration of conservation groups, private industry, and state and federal governmental agencies) has been formed under the auspices of the North American Waterfowl Management Plan (U.S. Fish and Wildlife Service 1988). This joint venture has increased recognition of the importance of playas and is the model for playa conservation efforts. Hopefully, these programs will be successful in future conservation of playas to protect the flora and fauna that depend on playas.

Without playa wetlands, biodiversity on the SHP would be extremely low. Additionally, the benefits of playas to humans such as aquifer recharge and flood storage are irreplaceable. Despite the unpredictable, stressful environment of playas and exploitation of uplands surrounding playas, many organisms have adapted and continue to thrive in these wetlands. We hope that the information in this book will not only lead to quicker and more accurate field identification of plant species in playas, but also to a better understanding of playa ecosystems.

SPECIES DESCRIPTIONS

Plants are arranged alphabetically by family. Within each family, species are ordered alphabetically by genus. In the species descriptions there are a number of categories that require some clarification. The common and scientific names of each species are primarily from the Great Plains Flora Association (1991). Where another name is listed in parentheses following the common name, it is a frequently used, local, colloquial term, or it was the common name listed in the database "National List of Plant Species that Occur in Wetlands" developed by the National Wetland Inventory (1996) of the U.S. Fish and Wildlife Service. Synonyms listed below the scientific name are other scientific names that have been used to designate the species. These were primarily obtained from the Great Plains Flora Association (1991) and Kartesz (1994). Synonyms designated by an asterisk indicate the appropriate scientific name according to Kartesz (1994).

The inflorescence and vegetative/life form sections contain descriptive material summarized with permission from the Great Plains Flora Association (1991) and from data collected during our surveys. Botanical terms are defined in the glossary. We included definitive floral measurements and attributes that are readily observable in the field. The growing season information is based on our survey results and published information, and is the period when one would expect to encounter the species in a playa.

The wetland indicator status was included to aid state and federal agencies or others who may be involved in regulatory activities impacting federal jurisdictional wetlands. We listed the National Wetland Inventory's wetland indicator status category for each species as determined by the rangewide frequency of occurrence in wetlands for each species in three regions: Region 5, Central Plains (Kansas, Colorado); Region 6, South Plains (Texas, Oklahoma); and Region 7, Southwest (New Mexico, Arizona). The categories were established based on the occurrence of species in wetlands in relation to their occurrences on surrounding uplands. Wetland indicator status is classified as: OBL,

Obligate Wetland, occurs with an estimated 99% probability in wetlands; FACW, Facultative Wetland, estimated 67–99% probability of occurrence in wetlands; FAC, Facultative, equally likely to occur in wetlands and nonwetlands (34–66% probability); FACU, Facultative Upland, 67–99% probability in nonwetlands, 1–33% in wetlands; UPL, Obligate Upland, more than 99% occurrence in nonwetlands; and NI, insufficient information available to determine indicator status. A listed "+" or "– " indicates a frequency towards the mesic (wetter) and xeric (dryer) end of the category, respectively. At times, there was no published indicator for species identified in playas. We addressed these by assigning a designation based only on our own vegetation surveys.

Relative abundance categories are based on our vegetation surveys. If a species was found in more than 25% of surveyed playas, it is classified as common; from 5–25% of playas, it is labeled as uncommon; and species in fewer than 5% of playas are listed as rare. Information on soil moisture conditions is based on our observations of the required hydrological conditions for germination and establishment of each species in playas. We group species into three categories based on conditions in which they were found: (1) wet/flooded, where standing water was found in playas; (2) moist, where the playa topsoil moisture was at or near saturation and considered "muddy;" and (3) dry, when topsoil in a playa was dry to the touch.

Under the habitat considerations section, we give any known value (positive or negative) of the plant to resident and migratory wildlife species. This includes food as well as cover values. If a species is actively managed for wildlife, some details are provided. Also, it is noted if a species is of particular forage value or poisonous to livestock.

We list the states and counties in which each species is likely to occur in playas. However, this information is based on our vegetation surveys and other published accounts of plant species in playas. For example, national distribution data of each species were used to indicate potential state occurrence. A designation of "all" counties refers to the counties containing playas in those states listed in the previous entry.

There are two plant species abundance categories: percentage of playas and percent total community composition. The "early" category for percentage of playas is simply the percentage of the 228 playas surveyed in the spring that contained that particular species. The "late" category is the percentage of the 229 playas surveyed in summer. The total category lists the percentage of a particular species occurring in all 235 surveyed playas. Because of the dynamic nature of plant communities in playas, it was possible for a species to be present in a playa during early sampling and disappear by the late sampling period or to become established in a playa following early sampling and thus be recorded only during late-season surveys. Percent total community composition is the total percent area (by step points) covered by a particular species in all playas during the respective sampling time (early, late). The total category for community composition lists the percent area covered by a species for early and late surveys combined. If a zero is listed in the community composition section, the species may have been present but it made up less than 0.001% of the community.

At the end of each species description, we list species that appear similar to the one being described and provide characteristics to separate the two in question. We only list similar species potentially found in playas based on our surveys and other published information.

FAMILY ALISMATACEAE

Longbarb Arrowhead (Duck Potato)
Sagittaria longiloba Engelm. *ex* J.G. Sm.

Species Description: *Inflorescence*—May reach 1.5 m tall, simple or with single branch at lowest node; bracts basally connate, ovate-lanceolate, attenuate, < 1.5 cm long; pedicels 0.7–3 cm long; lower whorls female, upper whorls male. Sepals 4–7 mm long; petals twice as long as sepals. Fruiting heads up to 1 cm wide; achenes obovate to oblong, 2.0–2.5 mm wide.

Vegetative/Life Form—Rhizomatous perennial with fall corms. Leaves erect or spreading, to 8 dm long. The only sagittate-leaved species with a consistent leaf shape. The narrow lobes and longer length of basal lobes up to 22 cm long distinguish this species.

Growing Season: Late spring through summer

Wetland Indicator Status:

	Region 5	Region 6	Region 7
	OBL	OBL	OBL

Abundance Category: Uncommon

Soil Moisture Conditions: Wet; germinates under a few cm of water. Usually becomes established after a playa has been shallowly flooded for > 2 weeks.

Habitat Considerations: This plant is considered a good waterfowl food as a result of the fall corms. It also provides fair cover in wet playas. To encourage this species, water depths of at least a few cm must be maintained from late spring through summer.

States: Only reported in Texas playas, possible in playas of Oklahoma and Kansas.

Counties: Usually found in eastern counties of the Texas portion of the PLR.

Percentage of Playas: Early 4.8, Late 12.3, Total 13.3

Percent Total Community Composition: Early 0.18, Late 1.00, Total 0.58

Similar Species: *Sagittaria calycina* (syn. *S. montevidensis*) has much wider leaves than *S. longiloba*.

FAMILY AMARANTHACEAE

Rough Pigweed (Red-root Amaranth)
Amaranthus retroflexus L.

Species Description: *Inflorescence*—Monoecious. Spikes terminal and/or axillary, paniculate, crowded, erect, 5–20 cm long, sepals of staminate flowers ovate-oblong to lanceolate, 3 mm long; pistillate flower sepals linear-oblong, 2.5–3.2 mm long; styles 3-branched. Seed shiny, round, black, about 1 mm in diameter.

Vegetative/Life Form—Annual herb with a red taproot. Stems erect, 0.3–3.0 m tall, simple to freely branched, roughish villous-puberulent, white or reddish striate. Leaves alternate, with lanceolate to obovate-oblanceolate blades 2–8 cm long; pubescent on lower surface; petioles about as long as blades.

Growing Season: Summer through fall

Wetland Indicator Status:	Region 5	Region 6	Region 7
	FACU	FACU–	FACU

Abundance Category: Common

Soil Moisture Conditions: Dry; germinates on exposed soils.

Habitat Considerations: Seeds commonly consumed by migratory and resident birds. Common weed in cultivated fields. Often considered good late-season brood cover for upland game birds. Sometimes promoted as a moist-soil plant in eastern U.S. but seldom promoted in playa management. Generally occurs along the edge of playas.

States: Reported from playas of Texas, New Mexico, Oklahoma, and Kansas; likely occurs in playas of Colorado.

Counties: All

Percentage of Playas: Early 13.2, Late 21.5, Total 29.6

Percent Total Community Composition: Early 0.34, Late 0.57, Total 0.46

Similar Species: Commonly hybridizes with other *Amaranthus* spp., which often makes it difficult to separate from other pigweeds.

FAMILY ASTERACEAE

Bur Ragweed (Lakeweed, Woolly-leaf Bursage)
Ambrosia grayi (A. Nels.) Shinners

Common Synonyms: *Franseria tomentosa, Gaertnera tomentosa, G. grayi*

Species Description: *Inflorescence*—Staminate inflorescence racemose-spicate; heads stalked, involucre ≤ 5 mm in diameter, 5–9 lobed. Pistillate heads in clusters or occurring singly in upper leaf axils. Involucre ≤ 7 mm long and ≤ 4 mm across.

Vegetative/Life Form—Herbaceous perennial 3–6 dm tall spreading by rootstalk to form large clonal groups. Leaves alternate, silvery gray, narrowed at the base to a petiole. Blades ovate to lanceolate, ≤ 10 cm long and ≤ 8 cm wide, irregularly pinnate-lobulate, lobes serrate.

Growing Season: Late spring through fall

Wetland Indicator Status:

	Region 5	Region 6	Region 7
	FAC	FACW	——

Abundance Category: Common; the most common species found in our surveys.

Soil Moisture Conditions: Dry to moist

Habitat Considerations: A noxious weed with little wildlife value. This species is discouraged by moist-soil management, which promotes annual seed-producing plants. Occurs throughout dry playas and along edges of wet playas.

States: All

Counties: All

Percentage of Playas: Early 70.5, Late 73.7, Total 76.0

Percent Total Community Composition: Early 12.50, Late 15.91, Total 14.21

Western Ragweed (Naked-spike Ragweed)
Ambrosia psilostachya DC.

Common Synonyms: *Ambrosia californica, A. coronopifolia,
A. cumanensis, A. rugelii*

Species Description: *Inflorescence*—Staminate heads in a
racemiform arrangement, oblique, 2.5 mm in diameter; pistillate
heads in axillary clusters below staminate flowers, 1-flowered;
involucre 2.5 mm long.

Vegetative/Life Form—Perennial, 3–6 dm tall, forming clonal groups
from rootstocks. Stems hirsute to pubescent with short, ascending
hairs, branching. Leaves subsessile, lanceolate to ovate (4–8 cm),
once pinnatifid, divisions linear-lanceolate to toothed.

Growing Season: Spring through fall

Wetland Indicator Status:	**Region 5**	**Region 6**	**Region 7**
	FAC	FACU	FAC

Abundance Category: Uncommon

Soil Moisture Conditions: Dry

Habitat Considerations: The species has some cover value and its
seeds are consumed by songbirds and game birds; however, seed
production is annually variable. Usually present only on playa
margins.

States: Reported in playas from all states but Colorado, where it is
likely to be present.

Counties: All

Percentage of Playas: Early 3.5, Late 3.5, Total 5.6

Percent Total Community Composition: Early 0.04, Late 0.05,
Total 0.05

Similar Species: May be confused with *Ambrosia grayi* (Bur Ragweed)
during seedling or early vegetative stage.

Saltmarsh Aster (Annual Aster)
Aster subulatus Michx.

Common Synonyms: *Aster exilis, A. inconspicuus, A. neomexicanus, A. sandwicensis*

Species Description: *Inflorescence*—Paniculate, diffuse inflorescence; involucre 5–8 mm tall; involucral bracts imbricated in 3 or 4 series, appressed, greenish towards apex, tip reddish purple; ligule purple, pink, or white, 3–4 mm long; disk florets with corolla yellowish.

Vegetative/Life Form—Glabrous annual, taprooted, 1–7 dm tall, with many upward branches. Leaves linear to linear-lanceolate, 1–10 cm long, 0.7–1.5 mm wide, entire or remotely serrate and sessile.

Growing Season: Early summer through fall

Wetland Indicator Status:	**Region 5**	**Region 6**	**Region 7**
	FACW	OBL	OBL

Abundance Category: Common

Soil Moisture Conditions: Moist to dry; germinates on exposed soils.

Habitat Considerations: May occur throughout a drying playa. Little known wildlife value. In moist-soil management situations this plant indicates more water should have been applied to encourage annual seed-producing plants.

States: Reported in playas from all states but likey occurs in Colorado.

Counties: All

Percentage of Playas: Early 15.9, Late 25.9, Total 27.0

Percent Total Community Composition: Early 0.58, Late 1.22, Total 0.91

Horse-weed (Mare's-tail)
Conyza canadensis (L.) Cronq.

Common Synonyms: *Erigeron canadensis, Leptilon canadense*

Species Description: *Inflorescence*—Elongate terminal cluster; involucre 3–4 mm tall; involucral bracts greenish, imbricate; pistillate florets in a single series, 20–40, ligule white. Achenes hirsute, pappus abundant, white.

Vegetative/Life Form—Annual 3–15 dm tall; stem erect, simple to the inflorescence; herbage coarse, spreading-hirsute, sometimes glabrescent. Leaves narrowly oblanceolate to linear, sessile, entire, lower ones sometimes coarsely toothed, cauline leaves ≤ 8 cm long and 1 cm wide.

Growing Season: Spring through fall

Wetland Indicator Status:	Region 5	Region 6	Region 7
	FACU–	UPL	FACU

Abundance Category: Common

Soil Moisture Conditions: Dry to moist; germinates on exposed soils.

Habitat Considerations: Usually found on the edge of playas but may be scattered throughout dry playas. Little wildlife value but may provide vertical cover. A weed in cultivated fields.

States: All

Counties: All

Percentage of Playas: Early 17.2, Late 16.2, Total 25.3

Percent Total Community Composition: Early 0.26, Late 0.23, Total 0.25

Plains Coreopsis
Coreopsis tinctoria Nutt.

Common Synonyms: *Coreopsis cardaminifolia, C. stenophylla*

Species Description: *Inflorescence*—Heads numerous, terminating on upper branches; ray florets with ligule 1.0–1.5 cm long, yellow usually with red spot at the base; outer involucral bracts linear-oblong to triangulate < 1/2 as long as inner involucral bracts. Achenes wingless.

Vegetative/Life Form—Annual, 6–12 dm tall. Stems single, much branched. Leaves subsessile or short-petiolate, 5–10 cm long, pinnate or bipinnate, segments linear to linear-lanceolate.

Growing Season: Spring through summer

Wetland Indicator Status:	**Region 5**	**Region 6**	**Region 7**
	FACU–	UPL	FACU

Abundance Category: Uncommon

Soil Moisture Conditions: Moist

Habitat Considerations: Often cultivated for flowers. Little known wildlife value, but certainly one of the more showy flowers occurring in playas.

States: Most common in playas of Texas, but may occur in the other states.

Counties: All in Texas, Oklahoma, and Kansas.

Percentage of Playas: Early 14.5, Late 12.3, Total 16.3

Percent Total Community Composition: Early 0.68, Late 0.42, Total 0.55

Curly-top Gumweed (Curly-cup Gumweed)
Grindelia squarrosa (Pursh) Dunal

Common Synonyms: *Grindelia perennis, G. serrulata*

Species Description: *Inflorescence*—Heads, disk 0.7–3 cm wide, numerous; involucral bracts strongly resinous, imbricate in several series; rays 12–37, 7–15 mm long, yellow. Achenes 2–3 mm long; pappus awns 2–8.

Vegetative/Life Form—Biennial, glabrous; stems 1-several from a herbaceous base, 1–10 dm tall. Leaves 1.5–7.0 cm long, 4–20 mm wide, ovate to linear-oblong, with callous-serrulate to coarsely toothed edges, punctate.

Growing Season: Spring through fall

Wetland Indicator Status:	**Region 5**	**Region 6**	**Region 7**
	FACU–	FACU–	FACU–

Abundance Category: Common

Soil Moisture Conditions: Dry

Habitat Considerations: Little known wildlife value. Occurs scattered through dry and moist playas and on edges of wet playas.

States: All

Counties: All

Percentage of Playas: Early 18.1, Late 21.5, Total 28.8

Percent Total Community Composition: Early 0.33, Late 0.37, Total 0.35

Similar Species: In vegetative form may be confused with *Haplopappus ciliatus* (Sawleaf Daisy) but *H. ciliatus* lacks the resinous feel of *G. squarrosa*.

Goldenweed (Sawleaf Daisy)
Haplopappus ciliatus (Nutt.) DC.

Common Synonyms: *Prionopsis ciliata* * (Nutt.) Nutt.

Species Description: *Inflorescence*—Heads few in an open cyme; involucre 2.0–2.5 cm wide; ray florets 25–30, yellow. Achenes, ellipsoid to oblong, 2–3 mm long; pappus of numerous bristles.

Vegetative/Life Form—Annual, up to 0.75 m tall, from a taproot. Leaves serrate, sessile, mostly on upper half of stem, oblong to obovate 3–5 cm long, 1–2 cm wide.

Growing Season: Summer through fall

Wetland Indicator Status:	Region 5	Region 6	Region 7
	UPL	FACU	FACU

Abundance Category: Uncommon

Soil Moisture Conditions: Dry

Habitat Considerations: Little known wildlife value; occurring mainly scattered on playa margins.

States: Reported in playas from Texas, New Mexico, and Colorado; likely in playas of Kansas and Oklahoma.

Counties: All

Percentage of Playas: Early 2.6, Late 5.3, Total 6.9

Percent Total Community Composition: Early 0.01, Late 0.06, Total 0.03

Similar Species: *Grindelia squarrosa* (Curly-top Gumweed)

38

Small-head Sneezeweed
Helenium microcephalum DC.

Common Synonyms: *Helenium ooclinium*

Species Description: *Inflorescence*—Heads with receptacle usually globose, yellow, ligules 3–6 mm long, disc florets reddish-brown. Achenes about 1 mm long.

Vegetative/Life Form—Taprooted annual. Stems simple below and branched above. Leaves narrowly elliptic, 3–8 cm long and 0.5–2.0 cm wide.

Growing Season: Summer through fall

Wetland Indicator Status:

	Region 5	Region 6	Region 7
	NI	FACW–	FACW

Abundance Category: Uncommon

Soil Moisture Conditions: Moist to dry

Habitat Considerations: Little known wildlife value; occurring throughout dry to moist playas.

States: Reported only in Texas playas.

Counties: Eastern counties of PLR in Texas: Briscoe, Carson, Crosby, Dawson, Floyd, Garza, and Lynn.

Percentage of Playas: Early 7.1, Late 6.6, Total 8.6

Percent Total Community Composition: Early 0.21, Late 0.21, Total 0.21

Similar Species: *Helenium amarum* (Bitter Sneezeweed) and *H. badium* have also been reported in Texas playas, but differ from *H. microcephalum* by having smooth and unwinged stems compared to the winged stem of *H. microcephalum*.

Common Sunflower
Helianthus annuus L.

Common Synonyms: *Helianthus aridus, H. lenticularis*

Species Description: *Inflorescence*—Heads on long peduncles; disk 2+ cm across, involucral bracts ovate to ovate-lanceolate, 3–5 mm wide, margin ciliate, ray florets ≥ 17, disk florets reddish to purple, chaffy bracts deeply 3-toothed. Achenes glabrate, 3–5 mm long.

Vegetative/Life Form—Taprooted annual, 6–25 dm tall, coarse, hairy, upwardly branched. Lowermost leaves opposite but most others alternate, long-petiolate, blade ovate to cordate, 10–40 cm long and about 1/2 as wide.

Growing Season: Late spring through fall

Wetland Indicator Status:	**Region 5**	**Region 6**	**Region 7**
	FACU	FAC	FAC–

Abundance Category: Common

Soil Moisture Conditions: Dry; germinates on exposed soils.

Habitat Considerations: Seeds consumed by northern bobwhites (*Colinus virginianus*), mourning doves (*Zenaida macroura*), ring-necked pheasants (*Phasianus colchicus*), waterfowl, and songbirds. Usually confined to playa margins.

States: All

Counties: All

Percentage of Playas: Early 10.6, Late 24.6, Total 27.9

Percent Total Community Composition: Early 0.14, Late 0.74, Total 0.44

Similar Species: *Helianthus petiolaris* (Plains Sunflower); both taprooted annuals but *H. annuus* has involucral bracts greater than 4 mm wide and *H. petiolaris* is white-hairy at the apex of central chaffy bracts. Hybrids between the two species occur throughout the PLR.

Texas Blueweed (Blueweed Sunflower)
Helianthus ciliaris DC.

Species Description: *Inflorescence*—Heads few, disk 1.5–2.5 cm across; ray florets 10–18, ligule about 1 cm long but often smaller; disk florets red, chaffy bracts, entire or 3-toothed. Achenes about 3 mm long.

Vegetative/Life Form—Perennial, glabrous, blue-green and glaucous in aspect, 3–7 dm tall. Stems arising from extensive, thin root stalks. Leaves opposite, lanceolate, with margin cileate, 3–7 cm long and 0.5–2 cm wide.

Growing Season: Late spring through summer

Wetland Indicator Status:	**Region 5**	**Region 6**	**Region 7**
	FAC	FAC	FAC

Abundance Category: Common

Soil Moisture Conditions: Dry to moist

Habitat Considerations: An aggressive weed in cultivated fields. Little wildlife value in playas. May occur throughout dry playas, but only occurs at the edges of wet playas. This plant is discouraged by moist-soil management.

States: Reported in playas of Texas, New Mexico, Colorado, and Oklahoma; likely occurs in playas of Kansas.

Counties: All

Percentage of Playas: Early 57.7, Late 51.3, Total 68.2

Percent Total Community Composition: Early 2.1, Late 1.5, Total 1.7

Bitterweed
Hymenoxys odorata DC.

Species Description: *Inflorescence*—Heads numerous, solitary on axillary or terminal peduncles; involucral bracts 4–6 mm long, outer bracts united at base, inner bracts exceeding outer bracts; ray florets 6–13. Achene 1.5–2.0 mm long.

Vegetative/Life Form—Taprooted, bushy-branched, aromatic annual, < 5 dm tall. Cauline leaves numerous, 2–10 cm long, pinnatisect into 3–15 filiform divisions.

Growing Season: Spring through early summer

Wetland Indicator Status:	**Region 5**	**Region 6**	**Region 7**
	NI	NI	NI

Abundance Category: Rare

Soil Moisture Conditions: Dry

Habitat Considerations: Little known wildlife value, but poisonous to livestock. Often considered an indicator of over-grazed range.

States: Reported from playas in Texas and New Mexico; likely in playas of Colorado, Kansas, and Oklahoma.

Counties: Reported in playas of the southwestern counties of Texas (Hockley, Terry) and Lea County, New Mexico; potentially found in all counties of the PLR.

Percentage of Playas: Early 3.5, Late 0.4, Total 3.4

Percent Total Community Composition: Early 0.02, Late 0, Total 0.01

Prickly Lettuce
Lactuca serriola **L.**

Common Synonyms: *Lactuca integrata, L. scariola, L. virosa*

Species Description: *Inflorescence*—A diffuse, conical panicle often > 50 heads. Heads cylindric; involucral bracts 17; florets 18–25; ligule yellow, 4–5 mm long, with blue stripe on abaxial side; corolla tube 3.0–3.5 mm long with hairs at the summit. Achene brown, 3 mm long, with 5–7 ridges.

Vegetative/Life Form—Herbaceous winter annual 0.5–1.5 m tall; stem with stiff bristles on lower 1/3, white latex sap. Lower cauline leaves sessile, obovate to ovate, dentate- to pinnate-lobed, base saggitate, 15 cm wide, abaxial midrib with bristles; upper cauline leaves lanceolate, dentate- to pinnate-lobed.

Growing Season: Spring through fall

Wetland Indicator Status:

	Region 5	Region 6	Region 7
	FAC	FAC	FAC

Abundance Category: Uncommon

Soil Moisture Conditions: Dry; germinates on exposed soils.

Habitat Considerations: Exotic. Little known wildlife value. Occurs mainly on playa margins but may be scattered throughout a dry playa.

States: All

Counties: All

Percentage of Playas: Early 11.5, Late 15.4, Total 19.7

Percent Total Community Composition: Early 0.16, Late 0.10, Total 0.13

Prairie Coneflower
Ratibida columnifera (Nutt.) Woot. & Standl.

Common Synonyms: *Lepachys columnifera, Ratibida columnaris, Rudbeckia columnaris, R. columifera*

Species Description: *Inflorescence*—Heads solitary or several at ends of long peduncles; columnar receptacle, ≤ 4.5 cm long; ligules yellow to purple, 1–3 cm long. Achenes 1–3 mm long; pappus of 1 or 2 teeth.

Vegetative/Life Form—Perennial 0.25–1.0 m tall from a taproot. Leaves alternate, pinnatifid to partly bipinnatifid, ≤ 15 cm long, 6 cm wide.

Growing Season: Summer through fall

Wetland Indicator Status: Not listed; likely FAC–.

Abundance Category: Uncommon

Soil Moisture Conditions: Dry to moist

Habitat Considerations: Little known wildlife value; usually occurs as individual plants on playa margins.

States: All

Counties: All

Percentage of Playas: Early 3.1, Late 3.9, Total 6.4

Percent Total Community Composition: Early 0.02, Late 0.02, Total 0.02

Similar Species: Vegetatively similar to *Ratibida tagetes* (Short-ray Prairie Coneflower), but *R. tagetes* has relatively smaller leaves and stems, and an inflorescence with a globular receptacle and pappus crown.

Short-ray Prairie Coneflower
Ratibida tagetes (James) Barnh.

Common Synonyms: *Lepachys tagetes, Rudbeckia globosa, R. tagetes*

Species Description: Inflorescence—Heads on naked peduncles; receptacle globular, 0.5–1.5 cm tall and 1 cm wide; ray florets, yellow to red. Achene oblong, 2–3 mm long; pappus a thick crown.

Vegetative/Life Form—Perennial, 15–40 cm tall, from taproot. Stems many-branched. Lower leaves lanceolate, entire to bipinnatifid, ≤ 13 cm long, upper leaves 3- to 5-cleft, < 30 mm long.

Growing Season: Early spring through summer

Wetland Indicator Status: Not listed; likely FAC–.

Abundance Category: Uncommon

Soil Moisture Conditions: Dry

Habitat Considerations: Little known wildlife value; occurs mainly throughout dry playas.

States: Reported in playas from Texas, New Mexico, Oklahoma, and Colorado; likely in playas of Kansas.

Counties: All in the PLR of New Mexico and Colorado; western counties of the PLR in Texas, Oklahoma, and Kansas.

Percentage of Playas: Early 7.5, Late 9.2, Total 13.7

Percent Total Community Composition: Early 0.20, Late 0.18, Total 0.19

Similar Species: *Ratibida columnifera* (Prairie Coneflower)

Goat's Beard (Salsify)
Tragopogon dubius Scop.

Common Synonyms: *Tragopogon major*

Species Description: *Inflorescence*—Corolla yellow. Heads solitary; involucral bracts typically 13. Achenes slender, 2.5–3.5 cm long, pappus a single series of white, plumose bristles.

Vegetative/Life Form—Biennial, milky juiced, from a taproot, to 0.75 m tall. Leaves alternate, ≤ 30 cm long, tapering from base to apex.

Growing Season: Spring through summer

Wetland Indicator Status: Not listed; likely FACU.

Abundance Category: Uncommon

Soil Moisture Conditions: Dry; germinates on exposed soils.

Habitat Considerations: Exotic. Little known wildlife value. Usually occurs as scattered individuals on playa margin.

States: All

Counties: All

Percentage of Playas: Early 8.8, Late 3.5, Total 10.7

Percent Total Community Composition: Early 0.03, Late 0.01, Total 0.02

Plains Ironweed
Vernonia marginata (Torr.) Raf.

Species Description: *Inflorescence*—Compact; heads with 8–26 florets, involucre 3.2–7.0 mm across; involucral bracts imbricate, purple/green, inner bracts 5.5–7.0 mm long, corollas 11–12 mm long. Pappus purple, bristles 8–10 mm long.

Vegetative/Life Form—Perennial herb 3–7 dm tall. Leaves numerous, short petiolate; middle cauline leaves broadly linear, glabrous, scabrous along veins above; pitted and punctate with glandular trichomes below.

Growing Season: Late spring through late summer

Wetland Indicator Status:	Region 5	Region 6	Region 7
	FAC	FAC	FACU

Abundance Category: Uncommon

Soil Moisture Conditions: Dry to moist; germinates on exposed soils.

Habitat Considerations: Little known wildlife value. Generally occurs scattered along the edge of moist and wet playas, sporadically throughout dry playas.

States: Only reported from playas in Texas and New Mexico; likely to occur in Oklahoma, Kansas, and Colorado.

Counties: Reported in playas of Armstrong, Bailey, Briscoe, Cochran, Floyd, Hale, Hockley, Lamb, Lubbock, Randall, and Terry, Texas; Curry and Lea, New Mexico.

Percentage of Playas: Early 5.7, Late 10.5, Total 12.9

Percent Total Community Composition: Early 0.04, Late 0.06, Total 0.05

Cocklebur
Xanthium strumarium L.

Common Synonyms: *Xanthium acerosum, X. chinense, X. commune, X. echinatum, X. glanduliferum, X. globosum, X. pensylvanicum, X. speciosum*

Species Description: *Inflorescence*—Heads in axillary clusters, unisexual, bur ovoid or cylindric 2.0–3.5 cm long, covered with stiff, hooked prickles and terminated by 2 incurved beaks.

Vegetative/Life Form—Coarse annual with taproot, 4–20 dm tall. Leaves, alternate long-petiolate, blades broadly ovate to suborbicular, ≤ 15 cm long, margin toothed, and often shallow-lobed.

Growing Season: Summer through fall

Wetland Indicator Status:	Region 5	Region 6	Region 7
	FAC	FAC–	FAC

Abundance Category: Uncommon

Soil Moisture Conditions: Dry; germinates on exposed soils.

Habitat Considerations: Considered a common weed in agricultural and moist-soil management situations. Usually found scattered on playa margins.

States: Reported from playas in Texas, New Mexico, Oklahoma, and Colorado; likely found in playas of Kansas.

Counties: All

Percent of Playas: Early 7.0, Late 7.5, Total 13.3

Percent Total Community Composition: Early 0.005, Late 0.06, Total 0.03

Similar Species: Can be confused with *Helianthus* spp. (Sunflower) in young vegetative stages, but floral characteristics are distinct.

FAMILY BRASSICACEAE

Peppergrass (Dense-flower Peppergrass)
Lepidium densiflorum **Schrad.**

Common Synonyms: *Lepidium bourgeauanum, L. Fletcheri,
L. neglectum, L. texanum*

Species Description: *Inflorescence*—Racemes numerous, erect,
5–10 cm long. Siliques oval to obovate, 2.0–3.5 mm long.

Vegetative/Life Form—Mostly annual, sometimes biennial, 2–5 dm
tall. Basal leaves oblanceolate to pinnatifid, 4–7 cm long.

Growing Season: Early spring through early summer

Wetland Indicator Status:	Region 5	Region 6	Region 7
	FAC	FAC	FAC

Abundance Category: Uncommon

Soil Moisture Conditions: Dry

Habitat Considerations: Little known wildlife value. Occurs mainly
on playa margins.

States: Reported in playas from Texas, Kansas, and Colorado; likely in
playas of New Mexico and Oklahoma.

Counties: All

Percentage of Playas: Early 10.1, Late 0.4, Total 9.9

Percent Total Community Composition: Early 0.08, Late 0, Total 0.04

Spreading Yellow Cress
Rorippa sinuata (Nutt.) A.S. Hitchc.

Common Synonyms: *Nasturtium sinuatum, Radicula sinuata*

Species Description: Inflorescence—Racemes axillary and terminal, 0.5–1.5 dm long; sepals 2.5–4.5 mm long, saccate; petals 4.0–5.5 mm long. Siliques short to elongate-cylindrical, 5.5–8.0 mm long; seeds 0.6–0.9 mm long, 20–60/silique.

Vegetative/Life Form—Perennial from rhizomes, stems decumbent, 1–4 dm long, pubescent with hemispherical vesicular trichomes. Middle cauline leaves sessile, oblong to oblanceolate, 3–8 cm long, 0.5–1.5 mm wide, sinuate, pinnatifid to subpinnatifid, the apex acute.

Growing Season: Early spring through mid-summer

Wetland Indicator Status:	Region 5	Region 6	Region 7
	FACW	FACW–	FACW

Abundance Category: Common

Soil Moisture Conditions: Moist

Habitat Considerations: Little known wildlife value; occurs sporadically, seldom in dense stands, throughout a moist playa.

States: All

Counties: All

Percentage of Playas: Early 41.9, Late 18.9, Total 46.4

Percent Total Community Composition: Early 0.78, Late 0.14, Total 0.45

Tumbling Mustard
Sisymbrium altissimum L.

Common Synonyms: *Norta altissima*

Species Description: *Inflorescence*—Flowers loosely racemose; petals yellow, 6–9 mm long. Siliques terete, linear, 5–10 cm long; seeds wingless, oblong, approximately 1 mm long.

Vegetative/Life Form—Annual, to 1 m tall, stems loosely branched above. Upper leaves linear-filiform entire segments, lower leaves petiolate, pinnately lobed, lobes oblong, dentate.

Growing Season: Spring

Wetland Indicator Status:	Region 5	Region 6	Region 7
	FACU+	FAC	FACU

Abundance Category: Uncommon

Soil Moisture Conditions: Dry; germinates on exposed soils.

Habitat Considerations: Exotic. Little known wildlife value. Occurs mainly on the margins of dry playas.

States: Reported in playas from Texas, New Mexico, Oklahoma, and Colorado; likely occurs in playas of Kansas.

Counties: All

Percentage of Playas: Early 6.2, Late 0, Total 6.0

Percent Total Community Composition: Early 0.04, Late 0, Total 0.02

Similar Species: *Descurainia pinnata* (Tansy Mustard) has branched hairs and much shorter siliques compared to the simple hairs and long siliques of *S. altissimum*.

FAMILY CAESALPINIACEAE

Indian Rush-pea (Hog Potato, Pignut, Waxy Rushpea)
Hoffmanseggia glauca (Ort.) Eifert

Common Synonyms: *Hoffmanseggia densiflora, Larrea densiflora*

Species Description: *Inflorescence*—Raceme glandular, pubescent, 1–2 dm long, 5- to 15-flowered. Petals orange-red, 10–13 mm long, with glandular claws; pedicels 2–5 mm, pubescent; calyx pubescent and glandular, lobes 6–7 mm. Legume indehiscent, flat, 2–4 cm long.

Vegetative/Life Form—Perennial, subscapose to caulescent; stems simple or branched, erect or spreading, 0.5–4.0 dm tall, from subterranean caudices from a root system with roundish tubers. Leaves subbasal or partly cauline, odd bipinnate; pinnae 2–6 pairs plus 1; leaflets 6–11 pairs, nearly sessile, oblong, 2.5–6.0 mm long, 2–3 mm wide.

Growing Season: Early spring through summer

Wetland Indicator Status:	Region 5	Region 6	Region 7
	FACU	FAC	FACU

Abundance Category: Uncommon

Soil Moisture Conditions: Dry to moist

Habitat Considerations: Noxious weed in several states. Underground tubers consumed by Native Americans. Habitat/food value for wildlife is unknown. Generally occurs along the edge of dry playas.

States: Reported in playas in Texas, New Mexico, and Oklahoma; possibly in playas of Kansas and Colorado.

Counties: Likely found in all PLR counties of Texas, New Mexico, and Oklahoma.

Percentage of Playas: Early 7.5, Late 6.6, Total 11.2

Percent Total Community Composition: Early 0.07, Late 0.04, Total 0.06

FAMILY CHENOPODIACEAE

Lamb's Quarters (White Goosefoot)
Chenopodium album L.

Common Synonyms: *Chenopodium amaranticolo*, *C. giganteum*, *C. lanceolatum*, *C. suecicum*

Species Description: *Inflorescence*—Glomerules clustered in dense, paniculate spikes. Sepals 5, farinose, median keel absent or not developed, enclosing the fruit at maturity; stamens 5, stigmas 2. Fruits horizontal, 1.1–1.5 mm in diameter.

Vegetative/Life Form—Annual, erect from a few cm to 1 m tall. Stems solitary, with lateral branches. Blades variable, narrowly trullate to lanceolate, > 1.5 times longer than wide, 3–5 cm long, 2–3 cm wide, farinose, acute.

Growing Season: Spring through fall

Wetland Indicator Status:	Region 5	Region 6	Region 7
	FAC	FAC	FAC–

Abundance Category: Common

Soil Moisture Conditions: Dry to moist; germinates on exposed soils.

Habitat Considerations: Exotic. Leaves and seeds may be consumed by resident and migratory wildlife. This species provides good brood cover for ring-necked pheasants (*Phasianus colchicus*) and northern bobwhites (*Colinus virginianis*). Can occur as solitary plants to dense stands in playas.

States: All

Counties: All

Percentage of Playas: Early 37.4, Late 39.0, Total 51.9

Percent Total Community Composition: Early 0.74, Late 0.99, Total 0.87

Similar Species: *Chenopodium leptophyllum* (Narrowleafed Goosefoot) has much narrower leaves (< 1.5 cm wide) with 1 vein from the base compared to the rounded (2–4 cm wide) leaves that have ≥ 3 veins from the base found in *C. album*.

Narrowleafed Goosefoot
Chenopodium leptophyllum (Moq.) Nutt. *ex* S. Wats.

Common Synonyms: *Chenopodium album* var. *leptophyllum*

Species Description: *Inflorescence*—Glomerules mostly spaced. Sepals 5, farinose. Fruits horizontal, about 1 mm in diameter, pericarp attached to seed.

Vegetative/Life Form—Annual to 9 dm tall. Stem solitary or infrequently branched at the base. Leaves alternate, linear blades, with single vein from the base.

Growing Season: Early summer through fall

Wetland Indicator Status:	**Region 5**	**Region 6**	**Region 7**
	FACU	FACU	FACU

Abundance Category: Common

Soil Moisture Conditions: Dry to moist; germinates on exposed soils.

Habitat Considerations: Little known wildlife value. Occurs mainly on margins of wet playas or sporadically throughout dry/moist playas.

States: Reported in playas from Texas, New Mexico, Oklahoma, and Colorado; likely occurs in playas of Kansas.

Counties: All

Percentage of Playas: Early 33.0, Late 41.7, Total 52.4

Percent Total Community Composition: Early 1.4, Late 1.1, Total 1.2

Similar Species: *Chenopodium album* (Lamb's Quarters)

70

Summer Cypress (Kochia, Fire-weed)
Kochia scoparia (L.) Schrad.

Common Synonyms: *Bassia sieversiana, Kochia alata, K. sieversiana, K. trichophila*

Species Description: *Inflorescence*—Long-spiciform to short, compact cylindric or oblong-claviform. Flowers perfect or functionally staminate or pistillate with some only pistillate on same plant, paired in axil, leaflike bracts 3–18 mm long. Fruit a utricle with persistent pericarp free from seed.

Vegetative/Life Form—Annual erect, 3–20 dm tall, branches erect or spreading. Stem yellow-green or green striped with red. Leaves alternate, 2–7 cm long, 0.5–8.0 mm wide, with 1 or 3 veins, lower leaves linear to lanceolate, distinct petiole, upper leaves linear, narrowly lanceolate, sessile base.

Growing Season: Late spring through early fall

Wetland Indicator Status:	**Region 5**	**Region 6**	**Region 7**
	FACU	FACU	FAC

Abundance Category: Common

Soil Moisture Conditions: Dry to moist; germinates on exposed soils.

Habitat Considerations: Good vertical cover value for upland game birds especially during brood-rearing and winter. Also provides cover for songbirds. Can be a good cattle forage. Occurs throughout dry playas and on margins of moist and wet playas.

States: All

Counties: All

Percentage of Playas: Early 47.1, Late 49.6, Total 61.8

Percent Total Community Composition: Early 3.5, Late 5.0, Total 4.2

Russian Thistle (Tumbleweed)
Salsola iberica Sennen & Pau

Common Synonyms: *Salsola kali, S. pestifer, S. tragus*

Species Description: *Inflorescence*—Short-spiciform, rigid, 1–7 cm long; bracts spreading and recurved, ovate to narrow deltoid, 3–8 mm long, apex spinose. Fruit obovoid with membranaceous, winged apex; seeds shiny, black, about 1.5 mm in diameter.

Vegetative/Life Form—Annual to 1 m tall; stems red-streaked, many branches without a main axis stem. Leaves alternate, filiform, 2–8 cm long, 1 mm wide, apex spinose, margins entire to denticulate.

Growing Season: Summer through fall

Wetland Indicator Status:	**Region 5**	**Region 6**	**Region 7**
	FACU	FACU	FACU

Abundance Category: Uncommon

Soil Moisture Conditions: Dry to moist; germinates on exposed soils.

Habitat Considerations: Exotic. Can provide good cover, especially during winter, for upland game birds and songbirds. Seeds are consumed by upland game birds and songbirds. Considered a weed in cultivated fields. Occurs mainly on playa margins.

States: All

Counties: All

Percentage of Playas: Early 10.1, Late 15.4, Total 20.2

Percent Total Community Composition: Early 0.04, Late 0.15, Total 0.09

74

Poison Suckleya
Suckleya suckleyana (Torr.) Rydb.

Species Description: *Inflorescence*—Flowers unisexual, in clusters in axils of leaf; staminate flowers above, calyx 3- to 4-parted, globose; pistillate flowers with 2 bracts. Fruit in bracts, pericarp membranaceous; seed red/brown, smooth, 2.8–3.1 mm long.

Vegetative/Life Form—Annual, stems succulent, 1–4 dm long. Leaves alternate; blades ovate to orbicular, 1–3 cm long, often green above and reddish below.

Growing Season: Late spring through summer

Wetland Indicator Status:	**Region 5**	**Region 6**	**Region 7**
	FACW	FACW	FACW

Abundance Category: Uncommon

Soil Moisture Conditions: Moist to dry; germinates on exposed soils.

Habitat Considerations: Little known wildlife value. May cause poisoning in livestock. A recent colonizer of playas.

States: Reported in playas of Texas and New Mexico; likely to occur in playas of Oklahoma, Kansas, and Colorado.

Counties: Reported in playas of Deaf Smith, Floyd, Hale, Hockley, Hutchinson, Lamb, Lubbock, Lynn, Parmer, and Randall, Texas; Curry and Lea, New Mexico.

Percentage of Playas: Early 4.8, Late 9.2, Total 11.6

Percent Total Community Composition: Early 0.24, Late 0.88, Total 0.56

FAMILY CUSCUTACEAE

Dodder
Cuscuta squamata **Engelm.**

Species Description: *Inflorescence*—Flowers sessile, separate or in clustered glomerules; calyx as long as the cylindric corolla tube. Fruits capsular, globose. Seeds 1–2, 1.5 mm long.

Vegetative/Life Form—Parasitic. Stems orange, ≤ 0.4 mm in diameter, vine. Leaves alternate, scale-like.

Growing Season: Summer through fall

Wetland Indicator Status: Not listed; likely FACU.

Abundance Category: Uncommon

Soil Moisture Conditions: Dry to moist

Habitat Considerations: Little known wildlife value. Growing on top of other plants (mainly Asteraceae) throughout dry to moist playas.

States: Reported in playas from Texas, New Mexico, and Kansas. Likely to occur in playas of Oklahoma and Colorado.

Counties: Primarily reported in central Texas and New Mexico counties of the PLR.

Percentage of Playas: Early 1.3, Late 8.3, Total 8.6

Percent Total Community Composition: Early 0, Late 0.06, Total 0.03

FAMILY CYPERACEAE

Umbrella Sedge (Chufa, Yellow Nutsedge)
Cyperus esculentus L.

Common Synonyms: *Cyperus lutescens*

Species Description: *Inflorescence*—Compound umbel of subsessile spike and 1–10 rays, involucral bracts broad and longer than the inflorescence; spikelets 5–20 mm long, 0.8–1.8 mm wide, 9–25/spike. Achenes trigonous, 0.9–1.5 mm long, 0.5–0.8 mm wide.

Vegetative/Life Form—Perennial rhizomatous, tubers on rhizome terminus. Culms triangular, 1–8 dm high. Leaves 3–10 mm wide, almost as long as the culm.

Growing Season: Late spring through summer

Wetland Indicator Status:	**Region 5**	**Region 6**	**Region 7**
	FACW	FACW	FACW

Abundance Category: Uncommon

Soil Moisture Conditions: Moist to shallowly flooded, spreading by tubers.

Habitat Considerations: May occur in scattered stands throughout a moist playa or on the edges of a wet playa. A preferred sandhill crane (*Grus canadensis*) food but not managed as such in playas.

States: Only reported in playas of Texas and New Mexico, but due to widespread nature of species likely in the other states as well.

Counties: All

Percentage of Playas: Early 3.1, Late 8.8, Total 10.3

Percent Total Community Composition: Early 0.02, Late 0.04, Total 0.03

Similar Species: In the vegetative form, other *Cyperus* spp.; they can only be separated based on floral characteristics.

Spikerush (Creeping Spikerush)
Eleocharis macrostachya Britt.

Common Synonyms: *Eleocharis palustris** (L.) Roemer &
J.A. Schultes, *E. mamillata, E. perlonga*

Species Description: *Inflorescence*—Spikelet solitary, terminal,
mainly lanceolate, ≤ 3 cm. Basal scales 2 or 3, sterile, fertile scales
purple to brown, ovate, narrow. Achenes yellow to brown, shiny,
obovate, ≤ 1.8 mm long.

Vegetative/Life Form—Perennial from reddish rhizomes. Culms green,
filiform ≤ 0.75 m, frequently flattened.

Growing Season: Spring through summer

Wetland Indicator Status:	Region 5	Region 6	Region 7
	OBL	OBL	OBL

Abundance Category: Common

Soil Moisture Conditions: Moist to shallowly flooded

Habitat Considerations: Where it occurs it may form dense stands
throughout the playa or it may occur along playa margins when
deeper water persists in the basin. Seeds consumed by migratory
birds, but production and nutritional quality of seed is relatively poor.
Therefore, other plants are preferred in moist-soil management plans.

States: All

Counties: All

Percentage of Playas: Early 61.7, Late 63.2, Total 72.1

Percent Total Community Composition: Early 15.13, Late 11.83,
Total 13.47

Similar Species: Several other *Eleocharis* spp. may occur in playas,
but *E. macrostachya* is the largest and most common spikerush in
playas. They are separated primarily by floral and seed characteristics.

Alkali Bulrush
Scirpus maritimus L.

Common Synonyms: *Bolboschoenus maritimus, B. paludosus, Schoenoplectus maritimus, Scirpus fernaldi, S. pacificus, S. paludosus*

Species Description: *Inflorescence*—Involucral bracts several, leaflike. Spikets many, ≤ 2 cm long, clustered, sessile, or peduncled. Achenes 3–4 mm long, lenticular or obovoid.

Vegetative/Life Form—Perennial with tuber-bearing rhizomes. Culms triangular in cross-section, ≤ 1.5 m tall. Leaves several, scattered on lower half of culm.

Growing Season: Spring through summer

Wetland Indicator Status:	Region 5	Region 6	Region 7
	OBL	OBL	OBL

Abundance Category: Rare

Soil Moisture Conditions: Moist to shallowly flooded, saline playas; germinates on moist soils.

Habitat Considerations: A good nesting substrate for overwater nesting ducks. Also a moderate seed producer, which is a good food source for waterfowl. Occurs on margins of wet, saline playas and can be found as scattered plants to dense stands.

States: Only reported in playas of Texas; likely found in playas of Kansas, Oklahoma, and New Mexico.

Counties: Primarily southern counties of the PLR in Texas containing saline semipermanent playas; Terry, Lynn, Dawson, and Howard counties.

Percentage of Playas: Early 0.88, Late 0.88, Total 0.88

Percent Total Community Composition: Early 0.03, Late 0.02, Total 0.03

Similar Species: *Scirpus fluviatilis* (River Bulrush), which only occurs in a very few northern playas, is usually larger.

83

Soft-stem Bulrush
Scirpus validus Vahl

Common Synonyms: *Scirpus tabernaemontani* * K.C. Gmel.,
Schoenoplectus validus, Schoenoplectus lacustris, Scirpus lacustris

Species Description: *Inflorescence*—Paniculate, spikelets single or in
glomerules of 2. Involucral bract culmlike. Scales exceeding achenes.
Achene 2.4 mm long.

Vegetative/Life Form—Perennial with large brown rhizomes. Culms
round, ≤ 3.5 m tall, light green.

Growing Season: Late spring through summer

Wetland Indicator Status:	Region 5	Region 6	Region 7
	OBL	OBL	OBL

Abundance Category: Uncommon

Soil Moisture Conditions: Shallow to deeply flooded for extended
periods of time; germinates mainly on moist soils.

Habitat Considerations: A good cover plant for wintering waterfowl
or for overwater nesting waterfowl. Seeds are consumed by
waterfowl, but seed production is generally low. Usually occurs in
scattered dense stands throughout a wet playa.

States: All

Counties: All

Percentage of Playas: Early 6.6, Late 10.5, Total 11.2

Percentage of Total Community Composition: Early 0.6, Late 1.0,
Total 0.8

Similar Species: Similar or same species (depending on the authority)
as *Scirpus acutus/S. lacustris* (Hardstem Bulrush) complex; floral and
vegetative characteristics all overlap.

FAMILY EUPHORBIACEAE

Snow-on-the-mountain
Euphorbia marginata Pursh

Common Synonyms: *Agaloma marginata, Dichrophyllom marginatum, Lepadena marginata*

Species Description: *Inflorescence*—Terminal, umbel-like, 3–4 rays, each ray branched; branches glabrate to woolly; floral bracts and leaves with white to pink margins. Cyathia solitary in branches of inflorescence; staminate flowers 35–60/cyathium. Capsule pubescent, 4–6 mm long.

Vegetative/Life Form—Annual, < 1 m tall, milky sap, unbranched below terminal inflorescence. Leaves alternate, sessile, whorled at base of umbel; oblong to elliptical, 3–10 cm long.

Growing Season: Summer through fall

Wetland Indicator Status:	**Region 5**	**Region 6**	**Region 7**
	FACU	FACU–	FACU–

Abundance Category: Uncommon

Soil Moisture Conditions: Dry

Habitat Considerations: Little known wildlife value. Occurs mainly scattered on playa margins; sometimes cultivated as an ornamental.

States: All

Counties: All

Percentage of Playas: Early 2.2, Late 11.0, Total 11.6

Percent Total Community Composition: Early 0.01, Late 0.07, Total 0.04

FAMILY FABACEAE

Yellow Sweet Clover
Melilotus officinalis **(L.) Lam.**

Species Description: *Inflorescence*—Flowers yellow, 4–7 mm long, spikelike racemes 4–15 cm long; peduncles axillary. Pod ovoid, usually 1-seeded, indehiscent.

Vegetative/Life Form—Annual to perennial, 0.5–2.0 m tall, with taproot. Leaves alternate, pinnately trifoliate; leaflets obovate to lanceolate, 1.0–2.5 cm long.

Growing Season: Spring through early fall

Wetland Indicator Status:	Region 5	Region 6	Region 7
	FACU	FACU	FACU+

Abundance Category: Uncommon

Soil Moisture Conditions: Dry

Habitat Considerations: Exotic. Considered a good cover for nesting and wintering ring-necked pheasants (*Phasianus colchicus*) and forage for large mammals. Often mixed with upland grass plantings, and hayed or grazed by cattle. Occurs mainly on playa margins.

States: Reported in playas from Texas and Colorado; likely occurs in playas of Kansas, Oklahoma, and New Mexico.

Counties: All

Percentage of Playas: Early 5.3, Late 3.5, Total 6.0

Percent Total Community Composition: Early 0.10, Late 0.06, Total 0.08

Similar Species: *Melilotus albus* (White Sweet Clover) is similar in vegetative form, but differs in flower color.

FAMILY GERANIACEAE

Filaria (Filaree, Stork's-bill)
Erodium cicutarium (L.) L' Hér *ex* Ait.

Common Synonyms: *Erodium aethiopicum*

Species Description: *Inflorescence*—Umbel long-peduncled, 2- to 8-flowered. Petals pink; sepals elliptic, 2–6 mm long. Seed ellipsoid, brown, 2–3 mm long.

Vegetative/Life Form—Winter annual, basal leaves from winter rosette, spreading branches to < 50 cm long. Leaves elongate, pinnate-pinnatifid.

Growing Season: Early spring through early summer

Wetland Indicator Status: Not listed; likely FAC.

Abundance Category: Rare

Soil Moisture Conditions: Dry to moist

Habitat Considerations: Green rosettes consumed by many species in winter including deer (*Odocoileus* spp.) and northern bobwhite (*Colinus virginianus*). Occurs on margins of moist playas.

States: Reported in playas of Texas.

Counties: Crosby, Garza, Hale, and Hockley.

Percentage of Playas: Early 1.8, Late 0, Total 1.7

Percent Total Community Composition: Early 0.003, Late 0, Total 0.002

Similar Species: *Erodium texanum* (Texas Stork's-bill) has simple leaf blades, whereas *E. filaria* has pinnate-pinnatifid leaf blades.

FAMILY LILIACEAE

False Garlic (Crow Poison)
Nothoscordum bivalve **(L.) Britt.**

Common Synonyms: *Allium bivalve, Ornithogalum bivalve*

Species Description: *Inflorescence*—Terminal umbel, from a spathe. Flowers light yellow, perianth of 6 segments, each with 1 nerve. Fruit a capsule, 3–5 mm long; seeds black.

Vegetative/Life Form—Scapose perennial 10–35 cm tall, from a bulb 1–2 cm thick. Leaves 2–6, basal linear, 2–5 mm wide.

Growing Season: Early spring through early summer

Wetland Indicator Status:	Region 5	Region 6	Region 7
	FACU	FACU–	——

Abundance Category: Uncommon

Soil Moisture Conditions: Dry

Habitat Considerations: Little known wildlife value. Usually occurs as scattered individual plants on playa margins.

States: Reported in playas of Texas.

Counties: Andrews, Armstrong, Bailey, Briscoe, Floyd, Garza, Hockley, Lynn, Swisher, and Terry.

Percentage of Playas: Early 7.5, Late 0.4, Total 7.7

Percent Total Community Composition: Early 0.02, Late 0, Total 0.01

Similar Species: Vegetatively resembles *Allium drummondii* (Wild Onion), but *N. bivalve* lacks an onion odor and has a yellow perianth, compared to the pink to white perianth of *A. drummondii*.

94

FAMILY LYTHRACEAE

Toothcup (Red-stem Ammania)
Ammannia auriculata Willd.

Species Description: *Inflorescence*—Flowers in axillary, simple or compound cymes, 3–12 flowers (commonly 7) per cyme, peduncles filiform, 3–9 mm long; petals 4, purple. Capsules 1.5–3 mm in diameter.

Vegetative/Life Form—Annual 1–8 dm tall, branches ascending when present. Leaves narrow, linear, the larger ones 17–64 mm long and 2–10 mm wide, auriculate and clasping at the base.

Growing Season: Summer through fall

Wetland Indicator Status:	**Region 5**	**Region 6**	**Region 7**
	OBL	OBL	OBL

Abundance Category: Rare

Soil Moisture Conditions: Occurs in playas that have been wet for > 1 month; requires moist soil or shallow flooding for germination.

Habitat Considerations: Occurs mainly on playa margins as scattered individuals. Little known wildlife value.

States: Reported in playas from all states but New Mexico and Colorado; likely occurring in these states as well given appropriate germination conditions.

Counties: All

Percentage of Playas: Early 0.4, Late 4.8, Total 5.2

Percent Total Community Composition: Early 0.002, Late 0.02, Total 0.01

Similar Species: *Ammannia coccinea* is a more robust species with 3–5 flowers per axil and capsules 3.5–5 mm in diameter compared to the typical ≥ 7 flowers per axil and capsules ≤ 2.5 mm in diameter of *A. auriculata*.

California Loosestrife
Lythrum californicum Torr. & Gray

Species Description: *Inflorescence*—Flowers solitary or paired in axils. Petals purple, 4–6 mm long; floral tubes 4–7 mm long.

Vegetative/Life Form—Perennial, somewhat woody, 20–80 cm tall, virgately branched. Leaves pale green, linear–oblong, 1.0–2.5 cm long, 3.0–8.0 mm wide, the lower stem leaves opposite to subopposite, the upper stem leaves mostly alternate.

Growing Season: Early summer through fall

Wetland Indicator Status:	Region 5	Region 6	Region 7
	OBL	OBL	OBL

Abundance Category: Uncommon

Soil Moisture Conditions: Moist

Habitat Considerations: Little known wildlife value. Occurs primarily on margins of moist playas as scattered individuals.

States: All

Counties: All

Percentage of Playas: Early 6.2, Late 16.7, Total 20.6

Percent Total Community Composition: Early 0.14, Late 0.42, Total 0.28

Similar Species: *Lythrum alatum* (Winged Loosestrife) may occur in northern playas, and differs from *L. californicum* qualitatively by being taller, with fewer, more robust stems, and darker, more ovate leaves.

FAMILY MALVACEAE

Cheeseweed (Alkali Mallow)
Malvella leprosa (Ortega) Krapov.

Common Synonyms: *Disella hederacea, Sida hederacea, S. leprosa*

Species Description: *Inflorescence*—Flowers solitary, axillary. Sepals united into an angular base; corolla white/cream to pink. Seeds three-angled, pendulous.

Vegetative/Life Form—Perennial, silvery, scurfy canescent with stellate hairs; stems branched, decumbent or prostrate, 5–40 cm long. Leaves asymmetrical, reniform to triangular-ovate, to 40 mm long, 50 mm wide.

Growing Season: Spring through early fall

Wetland Indicator Status:	Region 5	Region 6	Region 7
	FACW	FAC	FACW

Abundance Category: Common

Soil Moisture Conditions: Dry to moist

Habitat Considerations: Little known wildlife value. Occurs throughout a dry playa basin or on the margins of moist or wet playas.

States: All

Counties: All

Percentage of Playas: Early 64.3, Late 58.8, Total 70.4

Percent Total Community Composition: Early 5.04, Late 4.25, Total 4.64

Similar Species: *Malvella sagittifolia* (Narrow-leaved Cheeseweed), which has been reported in playas of eastern New Mexico, has narrowly triangular to nearly linear leaf blades compared to the wider-than-long leaf blades of *M. leprosa*.

FAMILY MARSILEACEAE

Western Water Clover (Pepperwort, Hairy Water Fern)
Marsilea vestita Hook. & Grev.

Common Synonyms: *Marsilia fournieri, M. minuta, M. mucronata, M. uncinata*

Species Description: *Inflorescence*—Sporocarps ellipsoid, on peduncles attached to the rhizomes, 5–8 mm long, 1.5–3.0 mm thick. Sporocarp pubescent, splitting at maturity into 2 valves and releasing 10–20 sori.

Vegetative/Life Form—Perennial, < 4 dm tall, creeping rhizomes. Leaves alternate, long-petiolate, 4-foliate.

Growing Season: Spring through fall

Wetland Indicator Status:	Region 5	Region 6	Region 7
	OBL	OBL	OBL

Abundance Category: Uncommon

Soil Moisture Conditions: Moist to shallowly flooded

Habitat Considerations: No known wildlife value. Usually occurs as scattered individuals on bare ground or in open water, but can be found in dense stands.

States: Has been reported in playas of Texas, New Mexico, Kansas, and Oklahoma; likely to be found in playas of Colorado.

Counties: All

Percentage of Playas: Early 9.7, Late 10.1, Total 15.9

Percent Total Community Composition: Early 0.20, Late 0.38, Total 0.29

FAMILY ONAGRACEAE

Spotted Evening Primrose
Oenothera canescens **Torr. & Frém**

Common Synonyms: *Gaurella canescens*

Species Description: *Inflorescence*—Flowers sessile, in axils; petals pinkish, spotted or streaked, 8–12 mm long Capsules 4-angled, beaked, 7–8 mm long; seeds brown, 0.9–1.1 mm long.

Vegetative/Life Form—Perennial in colonies, spreading by adventitious shoots; stems decumbent to ascending, branching from the base, 1–2 dm tall, strigulose to canescent. Leaves lanceolate, sinuate to mostly entire, 5–15 mm long, 2–7 mm wide.

Growing Season: Early spring through summer

Wetland Indicator Status:	**Region 5**	**Region 6**	**Region 7**
	FACW–	FAC	FAC

Abundance Category: Common

Soil Moisture Conditions: Dry to moist

Habitat Considerations: Little known wildlife value. Occurs mainly on playa margins.

States: All

Counties: All

Percentage of Playas: Early 60.8, Late 49.1, Total 67.8

Percent Total Community Composition: Early 1.43, Late 0.86, Total 1.14

FAMILY PEDALIACEAE

Devil's Claw (Unicorn Plant)
Proboscidea louisianica (P. Mill.) Thellung

Common Synonyms: *Martynia louisianica*

Species Description: *Inflorescence*—Flowers zygomorphic, corolla 5-lobed, white to purple. Fruit a capsule, bivalved and dehiscent, about 10–20 cm long, with an in-curved beak, splitting at maturity to form a 2-horned "claw."

Vegetative/Life Form—Annual with horse odor, up to approximately 1 m tall, with dense, articulate, glandular hairs. Leaves mainly opposite; blade reniform, 3–20 cm long, mostly ovate, veination palmate.

Growing Season: Summer through fall

Wetland Indicator Status:	Region 5	Region 6	Region 7
	FACU	FAC–	FAC

Abundance Category: Uncommon

Soil Moisture Conditions: Dry to moist

Habitat Considerations: Little known wildlife value; occurs scattered mainly on playa margins.

States: All

Counties: All

Percentage of Playas: Early 0.9, Late 7.0, Total 7.7

Percent Total Community Composition: Early 0.001, Late 0.02, Total 0.01

FAMILY POACEAE

Western Wheatgrass
Agropyron smithii Rydb.

Common Synonyms: *Pascopyrum smithii** (Rydb.) A. Löve,
Agropyron molle, A. palmeri, Elymus smithii, Elytrigia smithii

Species Description: *Inflorescence*—Spikes 3–16 cm long, some of
the nodes sometimes bearing 2 spikelets; spikelets 3–8 flowered;
glumes broadest below the middle, tapering to an acuminate or
awned tip, overall length 7–13 mm, the first slightly longer than the
second. Lemmas 7–15 mm long.

Vegetative/Life Form—Rhizomatous, glaucous, often glabrous
perennial 6–9 dm tall. Culms erect and hollow. Blades stiff, involute,
ridged above, 4–22 cm long and 1.0–5.5 mm wide.

Growing Season: Late spring through summer

Wetland Indicator Status:	Region 5	Region 6	Region 7
	FACU	FAC–	FAC–

Abundance Category: Common

Soil Moisture Conditions: Dry to moist

Habitat Considerations: Common forage grass for large mammals
and livestock. Can be fair nesting cover for songbirds and other
upland nesting birds if not heavily grazed. Occurs mainly on
playa margins or throughout the basin if the playa has been dry
for > 1 year.

States: All

Counties: All but southern counties of the PLR of Texas and New
Mexico.

Percentage of Playas: Early 28.2, Late 25.4, Total 33.1

Percent Total Community Composition: Early 4.4, Late 3.8, Total 4.0

Rescue Grass (Matua Prairie Grass)
Bromus unioloides Kunth

Common Synonyms: *Bromus catharticus** Vahl, *B. brevis,*
Ceratochloa cathartica, Festuca unioloides

Species Description: *Inflorescence*—Panicle usually contracted,
branches 3–13 cm long, often erect. Spikelets 4- to 10-flowered,
compressed–keeled, glabrous, 10–35 mm long, 3.5–11.0 mm wide;
first glume 5- to 7-nerved, 7.5–10 mm long; second glume
7- to 9-nerved, 8–12.5 mm long; lemmas keeled.

Vegetative/Life Form—Annual, cool-season, low-growing grass
1–4.5 dm tall; blades flat and usually pubescent.

Growing Season: Primarily late winter/early summer (February–June)

Wetland Indicator Status: None listed; FAC+

Abundance Category: Uncommon

Soil Moisture Conditions: Dry to moist; germinates on exposed soils.

Habitat Considerations: Often used as livestock forage grass.
Occurs on disturbed sediments usually throughout a playa. Little
value to migratory birds except as occasional forage for geese.
Probably also used by pronghorn (*Antilocapra americana*), cottontails
(*Sylvilagus* spp.), and other mammals as green forage.

States: Reported in playas of Texas, Oklahoma, and New Mexico;
likely to occur in playas of Colorado and Kansas.

Counties: All

Percentage of Playas: Early 11.9, Late 0.9, Total 12.0

Percent Total Community Composition: Early 0.38, Late 0.06,
Total 0.22

Similar Species: *Bromus japonicus* (Japanese Brome) has rounded
glumes and lemmas, compared to the strongly compressed-heeled
glumes and lemmas of *B. unioloides,* and is primarily found in
northern playas.

Buffalo Grass
Buchloe dactyloides (Nutt.) Engelm.

Common Synonyms: *Bulbilis dactyloides, Sesleria dactyloides*

Species Description: *Inflorescence*—Dioecious. Staminate flowers on slender culms, each with 2 or 3 unilateral spicate branches, each branch 7–11 mm long, spikelets in 2 rows; glumes unequal, 1-nerved. Pistillate spikelets in 2–3 burlike clusters, with 2–4 spikelets per bur, usually 1-flowered; first glume 0.5–4.5 mm long, membranaceous; second glume united with indurate axis of the bur, enveloping the lemma and terminating in 3 awnlike points.

Vegetative/Life Form—Perennial, stoloniferous, mat-forming, 0.3–2.0 dm tall. Culms solid. Blades flat, rolled in the bud; sheath open; ligule with short hairs < 0.9 mm long.

Growing Season: Usually spring through early summer but may produce until late summer depending upon precipitation.

Wetland Indicator Status:	Region 5	Region 6	Region 7
	FACU	FACU–	FACU

Abundance Category: Common

Soil Moisture Conditions: Dry to moist; germinates on exposed soils.

Habitat Considerations: Shortgrass species often dominating large areas. Important for livestock and other grazing mammals. Generally found on outer fringes of playa margins but may occur throughout the playa if it has been dry for > 1 year. Not dense enough to provide meaningful nesting cover.

States: All

Counties: All

Percentage of Playas: Early 47.6, Late 41.7, Total 54.5

Percent Total Community Composition: Early 5.4, Late 3.4, Total 4.5

Windmill Grass
Chloris verticillata Nutt.

Species Description: *Inflorescence*—Panicle, consisting of 6–20 branches, partly in verticels, along an axis 1–5 cm long. Spikelets widely spaced; first glume 1.7–2.5 mm long, the second 2.4–4.4 mm long; fertile lemma 2–3 mm long, awned from below the tip, the awn 4–9 mm long.

Vegetative/Life Form—Perennial 1.0–3.5 dm tall. Culms solid, erect above, decumbent and geniculate below; may root at lower nodes. Blades 1–12 cm long, 1.8–3.8 mm wide, glabrous to scabrous, folded in the bud, keeled toward the base.

Growing Season: Late spring through early fall

Wetland Indicator Status: Not listed; likely FACU.

Abundance Category: Rare

Soil Moisture Conditions: Dry

Habitat Considerations: Little known wildlife value; usually growing sparsely on playa margins.

States: All

Counties: All

Percentage of Playas: Early 2.2, Late 1.8, Total 3.9

Percent Total Community Composition: Early 0.03, Late 0.01, Total 0.02

Barnyard Grass
Echinochloa crusgalli (L.) Beauv.

Common Synonyms: *Echinochloa occidentalis, E. oryzicola, Panicum crusgalli*

Species Description: *Inflorescence*—Panicle, 5–27 cm long, primary branches appressed or spreading, bearing long setae which may be longer than the spikelets; first glume acute to acuminate 1–2 mm long, second glume 2.5–4.5 mm long; fertile lemma 2–4 mm long, acuminate, with a tip membranaceous and withering, differentiated from the body of the lemma, set off by tiny hairs.

Vegetative/Life Form—Annual, erect or prostrate 3–10 dm long. Culms glabrous. Blades and sheaths often with a few hairs on the margin of the collar region; blades 8–22 cm long, 6–12 mm wide; sheaths flat, compressed, keeled. Wide variation in characters under different environmental conditions.

Growing Season: Late spring through summer

Wetland Indicator Status:	**Region 5**	**Region 6**	**Region 7**
	FACW	FACW	FACW–

Abundance Category: Common

Soil Moisture Conditions: Moist; germinates on exposed soils.

Habitat Considerations: Excellent food for waterfowl and seed-eating resident birds. Typically a target species under moist-soil management regimes. Requires approximately 1.5–2 months to reach seed-bearing stage in playas. Can be a fair nesting cover if germinates in spring. Flooding with shallow water kills seedlings. Occurs throughout moist playas.

States: All

Counties: All

Percentage of Playas: Early 16.3, Late 48.7, Total 51.5

Percent Total Community Composition: Early 0.31, Late 4.64, Total 2.47

Similar Species: We include *Echinochloa muricata* under *E. crusgalli* because of similar floral and vegetative characteristics.

Prairie Cupgrass
Eriochloa contracta A.S. Hitchc.

Species Description: *Inflorescence*—Panicle, 5–11 cm long, spikelets in 2 rows and short-pediceled on one side of each principal branch; glumes unequal, the largest 5-nerved, pubescent, awned 3.2–4.5 mm long.

Vegetative/Life Form—Tufted annual 15–80 cm tall. Culms glabrous to short-pubescent, especially near the nodes. Blades rolled in the bud, flat at maturity, often short-pubescent, midrib not prominent, 4–17 cm long, 3–8 mm wide; ligule a dense fringe of hairs from a membranaceous base, 0.5–1 mm long.

Growing Season: Spring through fall

Wetland Indicator Status:	**Region 5**	**Region 6**	**Region 7**
	FACU	FAC+	FACW

Abundance Category: Not found in our surveys, but reported as common in cropped playas of Kansas.

Soil Moisture Conditions: Dry to moist

Habitat Considerations: Little known wildlife value.

States: Reported in playas of Kansas and northern Texas Panhandle.

Counties: Likely would be found in playas throughout the northern portions of the PLR.

118

Little Barley
Hordeum pusillum Nutt.

Common Synonyms: *Critesion pusillum*

Species Description: *Inflorescence*—Erect, stiff, spikelike, 2.5–7.0 cm long; glumes of central spikelet scabrous to pubescent, indistinctly 3-nerved, bowed at the base, lanceolate 3.5–5.5 mm long, tapering to a scabrous awn 2.5–7.0 mm long; glumes of the 2 lateral pedicellate spikelets dimorphic, the lowest glumes similar to those of central spikelet, the upper glumes setaceous, 5.0–11.0 mm long.

Vegetative/Life Form—Tufted or solitary annual, 1–4 dm tall. Culms generally erect, glabrous. Blades scabrous to pubescent, 1–10 cm long, 1.5–4.5 mm wide.

Growing Season: Spring to early summer

Wetland Indicator Status:	Region 5	Region 6	Region 7
	FAC	FACU	FAC

Abundance Category: Common

Soil Moisture Conditions: Dry to moist; germinates on exposed soils. Tolerates saline wetland conditions.

Habitat Considerations: Seeds commonly consumed by migratory birds, but it seldom occurs in sufficient densities to allow it to be a dominant portion of the avian diet. Also, because it generally occurs in sparse stands, it is not an important nesting cover. May occur throughout a dry playa.

States: All

Counties: All

Percentage of Playas: Early 32.2, Late 0.4, Total 31.3

Percent Total Community Composition: Early 0.94, Late 0, Total 0.47

Similar Species: *Hordeum jubatum* (Foxtail Barley) differs from *H. pusillum* as a perennial with awns > 2 cm.

120

Bearded Sprangletop
Leptochloa fascicularis (Lam.) Gray

Common Synonyms: *Diplachne acuminata, D. fascicularis, Leptochloa acuminata*

Species Description: *Inflorescence*—Panicle, 10–46 cm long; spikelets 7–11 mm long, 5- to 9-flowered; glumes unequal, the first 1.4–3.5 mm long, the second 2.5–4.6 mm long; lemmas bifid at tip, pubescent on lower portions of all 3 nerves.

Vegetative/Life Form—Tufted annual, 2–12 dm tall. Culms stout, erect to decumbent. Blades flat, scabrous, prominent white midrib, especially on upper surface near the base, 6–51 cm long, 2–6 mm wide.

Growing Season: Summer through fall

Wetland Indicator Status:	Region 5	Region 6	Region 7
	OBL	FACW	FACW

Abundance Category: Uncommon

Soil Moisture Conditions: Moist; germinates on exposed soils.

Habitat Considerations: May occur in disjunct stands throughout a playa. Small seeds likely consumed by songbirds, waterfowl, and small mammals. Provides good vertical cover in patches.

States: All

Counties: All

Percentage of Playas: Early 0.4, Late 9.2, Total 9.4

Percent Total Community Composition: Early 0, Late 0.06, Total 0.03

122

Common Witchgrass
Panicum capillare L.

Common Synonyms: *Panicum barbipulvinatum*

Species Description: *Inflorescence*—Panicles open, diffuse, purplish at maturity, 10–30 cm long; spikelets acuminate, 2–3 mm long; first glume acute, 3- to 7-nerved, 0.8–1.5 mm long.

Vegetative/Life Form—Tufted annual, approximately 0.5 m tall. Culms hollow, spreading, pubescent near the nodes. Blades flat, hispid on the margins, 6–16 mm wide, 5–20 cm long; sheaths papillose-hispid, not keeled.

Growing Season: Mid-summer through fall

Wetland Indicator Status:	Region 5	Region 6	Region 7
	FAC	FAC	FAC

Abundance Category: Uncommon

Soil Moisture Conditions: Dry to moist; germinates on exposed soils.

Habitat Considerations: Little known wildlife value; usually on the edge of dry playas.

States: All

Counties: All

Percentage of Playas: Early 0, Late 8.3, Total 8.2

Percent Total Community Composition: Early 0, Late 0.08, Total 0.04

Vine Mesquite
Panicum obtusum Kunth

Species Description: *Inflorescence*—Panicle narrow, 6–12 cm long, main branches appressed; spikelets blunt–tipped, 3.3–4.4 mm long; glumes subequal, 2.5–4.1 mm long, 5- to 9-nerved; lemma of lower floret about equal to glumes, usually 5-nerved; lemma of upper floret smooth, 2.6–3.6 mm long. Lower floret usually staminate.

Vegetative/Life Form—Perennial 2.5–7.0 dm tall, stolons from a knotty or short-rhizomatous base. Culms glabrous, hollow to solid, compressed to terete. Blades flat, nearly glabrous, 4–26 cm long, 3.0–6.3 mm wide; sheaths glabrous, except for a few hairs near the ligule; basal hairs villous especially at attachment to stolon.

Growing Season: Spring through summer.

Wetland Indicator Status:	Region 5	Region 6	Region 7
	FACW	FAC+	FAC

Abundance Category: Common

Soil Moisture Conditions: Dry to moist

Habitat Considerations: Used as a livestock forage, and probably consumed by other grazing mammals. When seeds are produced, they are consumed by waterfowl and other seed-eating birds. In dry/moist conditions, it will occur throughout the playa. Nesting cover value is limited.

States: Reported in playas of Texas, New Mexico, Oklahoma, and Kansas; possibly in playas of Colorado.

Counties: All

Percentage of Playas: Early 18.1, Late 15.8, Total 26.6

Percent Total Community Composition: Early 0.72, Late 0.48, Total 0.60

Similar Species: In vegetative form, *P. obtusum* can be confused with *Paspalum paspalodes* (Knotgrass), but inflorescences are very different.

Knotgrass (Joint Paspalum)
Paspalum paspalodes (Michx.) Scribn.

Common Synonyms: *Paspalum distichum* * L., *Digitaria paspaloides*

Species Description: *Inflorescence*—Two branches, 3–16 cm long, branch rachises 1.1–1.6 cm wide; spikelets single or paired, ovate, acute-tipped, 2.4–3.1 mm long; first glume present or absent; second glume thinly pubescent; sterile lemma and second glume, 3-nerved, equaling the spikelet.

Vegetative/Life Form—Stoloniferous, rhizomatous perennial. Culms solid, 4–8 dm tall, often decumbent below, compressed, and glabrous or pubescent at nodes. Blades flat, 3–22 cm long, 3–7 mm wide.

Growing Season: Summer through fall

Wetland Indicator Status:	**Region 5**	**Region 6**	**Region 7**
	OBL	FACW+	OBL

Abundance Category: Uncommon

Soil Moisture Conditions: Moist; germinates on exposed soils. Can survive extended flooded conditions after germination.

Habitat Considerations: Occurs world-wide in wetlands and can out-compete other wetland species. Seeds, when available, may be consumed by birds. Nesting cover value is fair. Promoted for livestock grazing in playas.

States: Reported only in playas of Texas; possible in playas of Oklahoma and New Mexico.

Counties: Briscoe, Castro, Cochran, Crosby, Dawson, Deaf Smith, Floyd, Garza, Gray, Hale, Hansford, Lamb, Lubbock, Lynn, and Swisher.

Percentage of Playas: Early 12.8, Late 18.4, Total 20.6

Percent Total Community Composition: Early 1.13, Late 2.22, Total 1.66

Similar Species: Sometimes confused with *Panicum obtusum* (Vine Mesquite) in its vegetative form.

May Grass (Carolina Canary Grass)
Phalaris caroliniana Walt.

Species Description: *Inflorescence*—Panicle cylindrical, 1–6 cm long; glumes winged, on the midnerve, ≤ 0.5 mm wide, pale with darker green on the wing and flanking the lateral nerves, 4–6 mm long; fertile lemma pilose, becoming shiny, acute, 2.9–4.1 mm long.

Vegetative/Life Form—Tufted or solitary-stemmed annual, 2–9 dm tall. Culms smooth. Blades glabrous to scabrous, 1–18 cm long; sheaths glabrous, air-chambered; ligules truncate, 2.5–5.0 mm long.

Growing Season: Early spring through mid-summer

Wetland Indicator Status:	**Region 5**	**Region 6**	**Region 7**
	FACW	FACW	FACW

Abundance Category: Uncommon

Soil Moisture Conditions: Moist; germinates on exposed soils.

Habitat Considerations: This is a recent invader of playa wetlands. Can occur throughout the playa. Its wildlife value is limited, but it may be a fair nesting cover. Residual cover following senescence may prevent germination of other species.

States: Reported from Texas playas.

Counties: Primarily central and eastern counties of Texas PLR: Crosby, Dawson, Floyd, Hansford, Lamb, Lubbock, Lynn, and Ochiltree.

Percentage of Playas: Early 8.8, Late 0, Total 8.6

Percent Total Community Composition: Early 0.85, Late 0, Total 0.42

130

Tumblegrass
Schedonnardus paniculatus (Nutt.) Trel.

Species Description: *Inflorescence*—Panicle with widely spaced spikes, the axis and branches curving when mature; spikelets appressed in 2 rows on 1 side of each branch, disarticulating above glumes, 1-flowered; glumes accuminate, 1-nerved; lemmas accuminate, 3-nerved.

Vegetative/Life Form—Tufted perennial, 1.5–5.0 dm tall. Culms solid, scabrous, flattened, curving above. Blades white-margined, keeled, folded in bud, V-shaped in cross-section, scabrous, especially on margins, 2–11 cm long, 1.0–2.8 mm wide; sheaths flat, keeled, open to the base.

Growing Season: Late spring through summer

Wetland Indicator Status: Not listed; likely FACU.

Abundance Category: Uncommon

Soil Moisture Conditions: Dry to moist

Habitat Considerations: Little known wildlife value; usually growing in sparse stands on playa margins.

States: All

Counties: All

Percentage of Playas: Early 10.6, Late 11.8, Total 16.3

Percent Total Community Composition: Early 0.14, Late 0.20, Total 0.17

Squirreltail (Bottlebrush Squirrel-tail)
Sitanion hystrix (Nutt.) J.G. Sm.

Common Synonyms: *Elymus elymoides * Raf., Elymus longifolius, Sitanion elymoides, S. longifolium*

Species Description: *Inflorescence*—Bilateral spike ≤ 20 cm long, including the awns, with 2 spikelets on alternating sides at each node; glume and awn together 2–9 cm long; spikelets 2- to 3-flowered.

Vegetative/Life Form—Tufted perennial up to 0.5 m tall. Culms hollow. Blades rolled in bud, flat or folded at maturity, ridged above, 2–4 mm wide.

Growing Season: Spring through early summer

Wetland Indicator Status:	**Region 5**	**Region 6**	**Region 7**
	FACU	FACU–	UPL

Abundance Category: Uncommon

Soil Moisture Conditions: Dry

Habitat Considerations: Little known wildlife value. Seeds may be consumed by songbirds and small mammals.

States: Reported in playas from Texas, Colorado, Oklahoma, and Kansas; likely in playas of New Mexico.

Counties: All in PLR of Colorado, Oklahoma, Kansas, and northern counties of Texas.

Percentage of Playas: Early 6.6, Late 2.6, Total 6.9

Percent Total Community Composition: Early 0.21, Late 0.01, Total 0.11

134

Johnson-grass
Sorghum halepense (L.) Pers.

Common Synonyms: *Holcus halepensis*

Species Description: *Inflorescence*—An open to contracted panicle 15–50 cm long with perfect sessile spikelets, each adjacent to a staminate pediceled spikelet; sessile spikelets dorsally compressed, glumes shiny, hardened; pediceled spikelets on pedicels 1.8–3.3 mm long, narrower than the sessile spikelets, glumes not hardened and more conspicuously nerved.

Vegetative/Life Form—Rhizomatous perennial 8–20 dm tall. Culms terete, solid, appressed, pubescent at nodes, elsewhere glabrous. Blades rolled in the bud, flat at maturity, midrib prominent, 10–90 cm long, 8–31 mm wide; ligule membranaceous with a prominent fringe.

Growing Season: Late spring through fall

Wetland Indicator Status:	Region 5	Region 6	Region 7
	FACU	FACU	FACU+

Abundance Category: Uncommon

Soil Moisture Conditions: Dry to moist; germinates on exposed soils.

Habitat Considerations: Exotic. Seeds are consumed by waterfowl and other seed-eating birds, but production is low compared to moist-soil plants, which may be hindered by its presence. Sandhill cranes (*Grus canadensis*) consume the rhizomes. Considered an aggressive weed in cultivated fields and may be poisonous to livestock in early growth stages. Usually found on playa margins.

States: Reported in playas of Texas; likely in playas of Colorado, New Mexico, Oklahoma, and Kansas.

Counties: All, but most common in playas of the southern portions of PLR in Texas.

Percentage of Playas: Early 14.9, Late 14.9, Total 22.3

Percent Total Community Composition: Early 0.36, Late 0.72, Total 0.54

FAMILY POLYGONACEAE

Water Smartweed
Polygonum amphibium L.

Common Synonyms: *Persicaria amphibia, P. coccinea, Polygonum coccineum, P. iowensis, P. pratincole*

Species Description: *Inflorescence*—Racemes 1–2, terminal. Perianth 4–5 mm long, red/pink. Achene lenticular, dark, shiny.

Vegetative/Life Form—Perennial rhizomatous, floating, prostrate to somewhat erect, forming dense mats. Leaves lanceolate to ovate, ≤ 25 cm long, 6 cm wide.

Growing Season: Spring through summer

Wetland Indicator Status:	**Region 5**	**Region 6**	**Region 7**
	OBL	OBL	OBL

Abundance Category: Uncommon

Soil Moisture Conditions: Wet to moist

Habitat Considerations: Usually considered a poor species in playas managed for wildlife because it forms dense mats with sparse and irregular seed production. The species occurs throughout a wet playa. Its perennial nature often makes it difficult to control.

States: Reported only from playas in Texas; likely to occur in playas of Kansas and Oklahoma.

Counties: All of the PLR of Texas.

Percentage of Playas: Early 11.5, Late 14.9, Total 18.0

Percent Total Community Composition: Early 1.66, Late 2.70, Total 2.17

Similar Species: May be confused with other *Polygonum* spp., but its prostrate perennial growth form and red/pink flowers distinguish it.

Pale Smartweed (Nodding Smartweed, Willow Smartweed)
Polygonum lapathifolium L.

Common Synonyms: *Persicaria incarnata, P. lapthifolia, Polygonum incarnatum, P. nodosum, P. oneillii, P. tomentosa*

Species Description: *Inflorescence*—Racemes numerous, drooping, in loose panicles at the stem ends, ≤ 7 cm long. Perianth white, 2–3 mm long; sepals 3-nerved, most nerves divided at the top. Achenes black, shiny, flat or biconcave.

Vegetative/Life Form—Annual from a taproot, up to 2 m tall. Leaves lanceolate, the tips acuminate, 5–20 cm long, 1–5 cm wide.

Growing Season: Early spring through summer

Wetland Indicator Status:	**Region 5**	**Region 6**	**Region 7**
	OBL	FACW–	OBL

Abundance Category: Uncommon

Soil Moisture Conditions: Moist to wet

Habitat Considerations: Important moist-soil plant that can be managed for waterfowl. Seed is consumed by migratory and resident birds. Occurs throughout a moist playa. Germinates on exposed soils in spring; if it receives 1–2 pulses of water in summer, it will form a dense stand with abundant seed.

States: Reported in playas of New Mexico and Texas; likely occurs in playas in Oklahoma, Kansas, and Colorado.

Counties: All

Percentage of Playas: Early 7.9, Late 12.7, Total 15.4

Percent Total Community Composition: Early 0.53, Late 1.16, Total 0.84

Similar Species: Vegetatively similar to *Polygonum pensylvanicum* (Pennsylvania Smartweed), but the midvein is divided at the top of the outer perianth in *P. lapathifolium* and the perianth is whitish, generally not pink as is *P. pensylvanicum*.

139

Pennsylvania Smartweed (Pink Smartweed)
Polygonum pensylvanicum L.

Common Synonyms: *Persica mississippiensis, Persicaria bicornis, P. longistyla, P. pensylvanica, Polygonum omissa*

Species Description: *Inflorescence*—Racemes spiciform, occurring throughout the plant. Perianth 3–5 mm long, rose or pink. Achene lenticular, shiny, with one or both sides concave.

Vegetative/Life Form—Erect annual, up to 2 m tall, with many branches and swollen nodes. Leaves lanceolate, to 15 cm long and 4 cm wide.

Growing Season: Spring through late summer

Wetland Indicator Status:	**Region 5**	**Region 6**	**Region 7**
	FACW+	FACW–	OBL

Abundance Category: Common

Soil Moisture Conditions: Moist to wet

Habitat Considerations: Important moist-soil plant that can be managed for waterfowl. The seed is consumed by resident and migratory birds. Occurs throughout a moist playa. Germinates on exposed soils in spring; if it receives 1–2 pulses of water in the summer, it will form a dense stand with abundant seed. Also forms good winter cover for ring-necked pheasants (*Phasianus colchicus*).

States: Reported in playas from Texas, Oklahoma, New Mexico, and Kansas; likely occurs in playas of Colorado.

Counties: All

Percentage of Playas: Early 39.7, Late 48.3, Total 54.9

Percent Total Community Composition: Early 2.33, Late 4.52, Total 3.56

Similar Species: We consider this the same species as *Polygonum bicorne*. Vegetatively similar to *P. lapathifolium* (Pale Smartweed).

Bushy Knotweed
Polygonum ramosissimum Michx.

Common Synonyms: *Polygonum allocarpum, P. autumnale, P. prolificum*

Species Description: *Inflorescence*—Inflorescence axillary, few-flowered. Perianth 5- to 6-lobed, 2–3 mm long. Achene dimorphic, brown or yellowish, 3-angular.

Vegetative/Life Form—Erect annual to 8 dm tall, from a taproot. Leaves numerous, blue-green, linear-lanceolate, 0.5–3.0 cm long; 1–5 mm wide.

Growing Season: Early summer through fall

Wetland Indicator Status:	**Region 5**	**Region 6**	**Region 7**
	FAC	FACW	FAC

Abundance Category: Uncommon

Soil Moisture Conditions: Dry to moist

Habitat Considerations: No known wildlife value; occurs as scattered individuals throughout dry and moist playas.

States: All

Counties: All

Percentage of Playas: Early 11.5, Late 15.4, Total 21.5

Percent Total Community Composition: Early 0.16, Late 0.15, Total 0.15

Similar Species: *P. ramosissimum* is distinguished from *P. aviculare* (Prostrate Knotweed) by being an erect species, whereas *P. aviculare* is usually prostrate or sprawling.

Pale Dock
Rumex altissimus Wood

Common Synonyms: *Rumex brittanicus, R. ellipticus*

Species Description: *Inflorescence*—Spiciform to 30 cm. Flowers imperfect; valves green to red/brown, 4–5 mm long. Achenes brown.

Vegetative/Life Form—Biennial/perennial erect up to 1 m tall, the stem ribbed and often branching below the inflorescence. Basal leaves long-petiolate, up to 20 cm long, 5 cm wide.

Growing Season: Early spring through summer

Wetland Indicator Status:	Region 5	Region 6	Region 7
	FAC	FACW+	FAC+

Abundance Category: Uncommon

Soil Moisture Conditions: Moist

Habitat Considerations: A prolific seed producer that can be found throughout a wet playa. Seeds are consumed by migratory and resident birds. Management practices to promote this plant are unknown.

States: Currently only reported in Texas playas; likely to occur in playas of New Mexico, Oklahoma, Kansas, and Colorado.

Counties: All

Percentage of Playas: Early 5.3, Late 8.3, Total 9.0

Percent Total Community Composition: Early 0.32, Late 0.32, Total 0.32

Similar Species: The leaves of *Rumex crispus* (Curly Dock) are much more distinctly crisped. *R. altissimus* is vegetatively similar to *Polygonum amphibium* (Water Smartweed).

146

Curly Dock
Rumex crispus L.

Species Description: *Inflorescence*—Racemes in a panicle, large, to 30–40 cm long. Achenes red/brown, abundant. Flowers perfect; valves 4–5 mm long, ovate to deltoid.

Vegetative/Life Form—Biennial/perennial, erect to approximately 1 m tall. Stems simple below the inflorescence. Leaves irregularly crisped; basal leaves a rosette-like form, to 30 cm long and 5 cm wide.

Growing Season: Early spring through mid-summer. Germination primarily occurs in late summer and fall, and the plant overwinters as a rosette.

Wetland Indicator Status:	Region 5	Region 6	Region 7
	FACW	FACW	FACW

Abundance Category: Common

Soil Moisture Conditions: Moist

Habitat Considerations: Exotic. A good moist-soil plant for waterfowl and resident birds; however, at present it is difficult to predict its abundance. Fall moisture seems to promote germination; then it flowers and dies the following summer. A good seed producer that may be found throughout a moist playa.

States: All

Counties: All

Percentage of Playas: Early 35.2, Late 30.3, Total 41.6

Percent Total Community Composition: Early 1.21, Late 1.04, Total 1.12

Similar Species: See *Rumex altissimus* (Pale Dock). *R. obtusifolius* (Bitter Dock) may hybridize with *R. crispus,* but the leaves are only slightly crisped compared to the heavily crisped leaves of *R. crispus.*

FAMILY PONTEDERIACEAE

Blue Mud-plantain
Heteranthera limosa **(Sw.) Willd.**

Common Synonyms: *Pontederia limosa*

Species Description: *Inflorescence*—Pedunculate, sheathing spathe 2–4 cm long, enclosing a single flower. Flower 1–3 cm wide, white or blue. Fruit a multiple-seeded capsule.

Vegetative/Life Form—Stems elongate when submersed or contracted when emergent. Leaves erect or blades floating, lancelolate to ovate.

Growing Season: Late spring through summer

Wetland Indicator Status:	**Region 5**	**Region 6**	**Region 7**
	OBL	OBL	OBL

Abundance Category: Uncommon

Soil Moisture Conditions: Wet/shallow flooded conditions; germinates under shallowly flooded conditions.

Habitat Considerations: Little known wildlife value. Generally occurs throughout shallowly flooded playas. A playa needs to be flooded for > 1 month before this species becomes established.

States: Reported in playas of Texas and Kansas; likely occurs in other states, given appropriate germination conditions.

Counties: All

Percentage of Playas: Early 1.3, Late 6.1, Total 7.3

Percent Total Community Composition: Early 0.002, Late 0.25, Total 0.12

FAMILY PORTULACACEAE

Common Purslane
Portulaca oleracea **L.**

Common Synonyms: *Portulaca neglecta, P. retusa*

Species Description: *Inflorescence*—Flowers yellow, sessile, solitary or in terminal glomerules, 5–10 mm wide. Capsules 5–9 mm tall; seeds black, granulate.

Vegetative/Life Form—Annual; stems fleshy, purple/red, prostrate. Leaves alternate, succulent, flat, spatulate to obovate, 1–3 cm long, 0.2–13.0 mm wide.

Growing Season: Late spring through fall

Wetland Indicator Status:	Region 5	Region 6	Region 7
	FAC	FAC	FAC

Abundance Category: Uncommon

Soil Moisture Conditions: Dry

Habitat Considerations: Exotic, with no known wildlife value. Occurs sporadically throughout a dry playa or on edges of moist playa.

States: Reported in playas of Texas, New Mexico, and Kansas; likely occurs in playas of Colorado and Oklahoma.

Counties: All

Percentage of Playas: Early 4.4, Late 10.1, Total 12.9

Percent Total Community Composition: Early 0.02, Late 0.30, Total 0.16

Similar Species: *Portulaca mundula* (Shaggy Portulaca) also occurs in dry playas, but differs from *P. oleracea* by having leaf axils that are conspicuously villous with long white hairs.

FAMILY POTAMOGETONACEAE

Longleaf Pondweed
Potamogeton nodosus Poir.

Common Synonyms: *Potamogeton americanus, P. fluitans*

Species Description: *Inflorescence*—Spikes dense, cylindric, 2–6 cm long. Fruits red/brown, 2.7–4.3 mm long, sharp dorsal keel.

Vegetative/Life Form—Perennial from rhizomes and/or tubers. Stems subterete, 1–2 mm thick, rarely branched, ≤ 1.5 m long. Leaves elliptic to oblong-elliptic, floating, 5–13 cm long.

Growing Season: Early summer through fall

Wetland Indicator Status:	Region 5	Region 6	Region 7
	OBL	OBL	OBL

Abundance Category: Rare

Soil Moisture Conditions: Wet/flooded

Habitat Considerations: Consumed by herbivorous waterfowl and provides good aquatic invertebrate substrate. Germinates under submersed conditions. One of the few floating-leaved aquatics in playas. Requires extended flooded conditions to become established.

States: Reported in playas of Texas; likely to occur in playas of Kansas and Oklahoma.

Counties: Primarily eastern counties of PLR, where precipitation is greatest.

Percentage of Playas: Early 0, Late 1.3, Total 1.3

Percent Total Community Composition: Early 0, Late 0.15, Total 0.07

FAMILY SALICACEAE

Black Willow
Salix nigra Marsh.

Common Synonyms: *Salix ambigua, S. denudata, S. dubia, S. falcata, S. flavovirens, S. gooddingii, S. ligustrina, S. ludoviciana, S. purshiana*

Species Description: *Inflorescence*—Catkins emerge with leaves in spring; pistillate catkins 4–10 cm long; bracts caducous, yellow, pubescent. Capsules ovoid, 3–5 mm long. Fruits in May/June.

Vegetative/Life Form—Tree to 20 m tall, generally < 10 m in playas; twigs red to gray/brown; branchlets yellow to red/brown. Leaves the same shade of green on both sides, lanceolate, 4–15 cm long, 7–20 mm wide.

Growing Season: Spring through summer

Wetland Indicator Status:	Region 5	Region 6	Region 7
	OBL	FACW+	——

Abundance Category: Uncommon

Soil Moisture Conditions: Dry to wet

Habitat Considerations: Generally only occurs on margins of playas modified for water retention, road ditches, etc. Important nesting habitat for mourning doves (*Zenaida macroura*), woody habitat for songbirds, perches for raptors, and winter cover for ring-necked pheasants (*Phasianus colchicus*). Predominant woody species found in playas.

States: Reported in playas from Texas, but may occur in Oklahoma and Kansas playas as well.

Counties: Primarily central and southern counties of PLR in Texas.

Percentage of Playas: Early 6.6, Late 7.9, Total 7.7

Percent Total Community Composition: Early 0.12, Late 0.13, Total 0.12

Similar Species: *Salix amygdaloides* (Peachleaf Willow) and *S. exigua* (Sandbar Willow) have been reported in playas. Leaves of *S. amygdaloides* pale to white-glaucous beneath, compared to the green on both sides of leaves from *S. nigra*. *S. exigua* is shrub, forming dense thickets, compared to the tree life-form of *S. nigra.*

FAMILY SCROPHULARIACEAE

Water Hyssop
Bacopa rotundifolia (Michx.) Wettst.

Common Synonyms: *Bacopa nobsiana, B. simulans, Bramia rotundifolia, Hydranthelium rotundifolium, Macuillamia rotundifolia*

Species Description: *Inflorescence*—1 to 4 flowers in axils of upper leaves. Calyx 5-parted, segments 3.0–4.5 mm long. Corolla white with yellow throat. Capsule subglobose 3.5–5.5 mm long, with numerous small seeds.

Vegetative/Life Form—Stems floating or prostrate, rooting at lower nodes. Leaves opposite, sessile, entire, palmately veined, obovate to suborbicular, 12–27 mm long, 12–23 mm wide.

Growing Season: Summer through fall

Wetland Indicator Status:	**Region 5**	**Region 6**	**Region 7**
	OBL	OBL	OBL

Abundance Category: Rare

Soil Moisture Conditions: Wet/shallowly flooded to moist; germinates under submersed conditions.

Habitat Considerations: No known wildlife value.

States: Reported in playas of Texas and Kansas; likely occurs in playas of Oklahoma.

Counties: Primarily eastern counties of PLR where precipitation is greatest.

Percentage of Playas: Early 0, Late 1.3, Total 1.3

Percent Total Community Composition: Early 0, Late 0.003, Total 0.002

FAMILY SOLANACEAE

Purple Ground Cherry (Prairie Lantern)
Quincula lobata (Torr.) Raf.

Common Synonyms: *Physalis lobata*

Species Description: *Inflorescence*—Peduncles commonly in pairs from leaf axils, flowers regular; corolla rotate, purple, 1.5–2.0 cm wide. Fruiting calyx 5–lobed; berry ovoid, green/yellow 5–8 mm in diameter; seed brownish, irregular, somewhat flattened, about 2 mm long.

Vegetative/Life Form—Perennial rhizomatous, decumbent, spreading from many branches at the base, ≤ 10 cm tall. Leaves alternate, somewhat fleshy; blades oblanceolate to linear-lanceolate, main leaves 4–10 cm long, 0.5–3 cm wide.

Growing Season: Early spring through early summer

Wetland Indicator Status:	Region 5	Region 6	Region 7
	NI	NI	NI

Abundance Category: Uncommon

Soil Moisture Conditions: Dry

Habitat Considerations: Little known wildlife value; occurs mainly on playa margins.

States: All

Counties: All

Percentage of Playas: Early 6.2, Late 0.4, Total 6.0

Percent Total Community Composition: Early 0.01, Late 0.003, Total 0.008

Similar Species: *Physalis viscosa* (Yellow Ground Cherry), which has been reported in southern playas of Texas and New Mexico, has a nodding, yellow flower compared to the erect, purple flower of *Q. lobata*.

Silver-leaf Nightshade (White Horse-nettle)
Solanum elaeagnifolium Cav.

Species Description: *Inflorescence*—Cymose or racemose, axillary near ends of branches. Calyx 5-angled; corolla pale blue, 20–30 mm wide. Berry yellow, turning black in fall and winter; seeds brown, ovoid, smooth, 3–5 mm long.

Vegetative/Life Form—Perennial with extensive rootstocks, to 0.75 m tall, silvery with dense pubescence of stellate hairs; stems with sharp, orange spines. Leaves alternate, blades lanceolate to oblong, 3–10 cm long.

Growing Season: Spring through summer

Wetland Indicator Status: Not listed; likely FACU.

Abundance Category: Common

Soil Moisture Conditions: Dry

Habitat Considerations: Little known wildlife value; usually found on playa margins. A serious weed in cultivated fields throughout the region.

States: All

Counties: All

Percentage of Playas: Early 29.5, Late 24.1, Total 40.3

Percent Total Community Composition: Early 0.24, Late 0.21, Total 0.22

FAMILY SOLANACEAE

Buffalo Bur (Kansas Thistle)
Solanum rostratum Dunal

Common Synonyms: *Androcera rostrata, Solanum cornutum*

Species Description: *Inflorescence*—Corolla yellow, 1.5–2.5 cm wide, lobes triangular. Flowers several to 20. Berry about 1 cm long, enclosed by spiny, enlarged calyx; seeds dark brown/black, pitted, about 2.5 mm long.

Vegetative/Life Form—Annual with taproots; stems up to 0.75 m tall, branching, pubescent, with numerous yellow spines. Leaves alternate, obovate/elliptic, once or twice pinnately lobed, surfaces with spines.

Growing Season: Late spring through early fall

Wetland Indicator Status: Not listed; likely FACU.

Abundance Category: Uncommon

Soil Moisture Conditions: Dry; germinates on exposed soils.

Habitat Considerations: Little known wildlife value. Occurs mainly on the edges of playas. Possibly poisonous to hogs.

States: All

Counties: All

Percentage of Playas: Early 4.4, Late 16.2, Total 18.0

Percent Total Community Composition: Early 0.01, Late 0.24, Total 0.13

FAMILY TYPHACEAE

Southern Cattail (Narrow-leaved Cattail)
Typha domingensis (Pers.)

Common Synonyms: *Typha angustata*

Species Description: *Inflorescence*—Stout spike, monoecious. Flowers very small, densely arranged, without a perianth; pistillate flowers separate and below staminate flowers on the spike.

Vegetative/Life Form—Perennial. Leaves long, green, to 1.5 cm wide and 2.5 m tall, arising from stout rhizomes.

Growing Season: Spring through summer

Wetland Indicator Status:

	Region 5	Region 6	Region 7
	OBL	OBL	OBL

Abundance Category: Uncommon

Soil Moisture Conditions: Wet to moist; an indicator of prolonged moist/flooded conditions. Germinates under very shallow (< 5 cm) water or in moist soil.

Habitat Considerations: Good winter cover for ring-necked pheasants (*Phasianus colchicus*) and other wildlife such as barn owls (*Tyto alba*). However, it is poor nesting cover and can choke out more desirable food and nest cover plant species when it occurs in dense stands.

States: All

Counties: All

Percentage of Playas: Early 6.6, Late 10.5, Total 10.7

Percent Total Community Composition: Early 1.2, Late 1.3, Total 1.2

Similar Species: *Typha* spp. (Cattail) taxonomy is complex, with most species hybridizing. *T. latifolia* (Broad-leaved Cattail) staminate and pistillate flowers meet on the spike, whereas in *T. domingensis* and *T. angustifolia* (Narrow-leaved Cattails), staminate and pistillate flowers are separate on the spike. Staminate portion of spike of *T. domingensis* in playas disappears by late summer.

FAMILY VERBENACEAE

Frog-fruit
Lippia nodiflora (L.) Michx.

Common Synonyms: *Phyla nodiflora* * (L.) Greene, *Lippia canescens, L. incisa, L. reptans*

Species Description: *Inflorescence*—Flowers in dense, pedunculate, axillary spikes. Spike 5–8 mm in diameter; corolla zygomorphic, pink/purple to white. Nutlets 2, ovoid, about 1 mm long.

Vegetative/Life Form—Perennial; stems 4-angled, prostrate, rooting at the nodes. Leaf blades spatulate to oblanceolate, more rounded at the apex; larger ones with several teeth on each side.

Growing Season: Late spring through fall

Wetland Indicator Status:	**Region 5**	**Region 6**	**Region 7**
	FACW	FAC	FACW

Abundance Category: Common

Soil Moisture Conditions: Dry to moist

Habitat Considerations: Little known wildlife value, but is one of the most ubiquitous species in playas. Occurs on edges of wet playas and throughout dry playas.

States: All

Counties: All

Percentage of Playas: Early 31.7, Late 32.0, Total 43.8

Percent Total Community Composition: Early 0.92, Late 0.87, Total 0.89

Similar Species: *Lippia cuneifolia* (Wedgeleaf Frog-fruit) is closely related and also reported in playas. *L. nodiflora* has wider leaves (6–25 mm wide) and often roots at nodes, compared to *L. cuneifolia* which has narrower leaves (2–8 mm wide) and infrequently roots at nodes.

168

Prostrate Vervain
Verbena bracteata Lag. & Rodr.

Common Synonyms: *Verbena bracteosa, V. imbricata*

Species Description: *Inflorescence*—Spikes terminal, sessile, 2–20 cm long. Calyx 3–4 mm long; corolla purplish, tube slightly longer than calyx. Nutlets yellow to brown, about 2.5 mm long.

Vegetative/Life Form—Mostly annual or sometimes short-lived perennial in playas. Stems several from the base, many branches, prostrate or decumbent, 1–5 dm long. Leaves opposite, lanceolate, pinnately incised or 3-lobed, 1–7 cm long, 0.5–3.0 cm wide.

Growing Season: Spring through fall

Wetland Indicator Status:	**Region 5**	**Region 6**	**Region 7**
	FACU	FAC	FAC

Abundance Category: Common

Soil Moisture Conditions: Dry to moist

Habitat Considerations: Little known wildlife value; generally occurs on playa margins.

States: All

Counties: All

Percentage of Playas: Early 17.2, Late 13.2, Total 25.3

Percent Total Community Composition: Early 0.13, Late 0.15, Total 0.14

APPENDIX
LIST OF PLANT SPECIES
OCCURRING IN PLAYA LAKES

For an explanation of abbreviations used for wetland indicator status see page 17.
Sources: (a) Our 1995 field study; (b) Colorado, Hoagland (1991) report; (c) Rowell
(1981) includes information from Rowell (1971); (d) Curtis and Beierman (1980); (e) Reed
(1930); (f) Cushing et al. (1993), Johnston (1995); (g) Kansas; Kindscher and Lauver
(1993), Kindscher (1994), Natural Resources Conservation Service (unpublished data).

Family/Species	Common Name	Wetland Indicator Status			Source
		Region 5	Region 6	Region 7	
Aizoaceae					
Sesuvium portulacastrum	Purslane Sesuvium	——	FACW	——	d
Alismataceae					
Alisma subcordatum	Water Plantain	OBL	OBL	O3L	f
Echinodorus rostratus	Burhead	OBL	OBL	OBL	acf
Sagittaria calycina	Arrowhead	OBL	OBL	OBL	af
Sagittaria cuneata	Arrowhead	OBL	OBL	OBL	de
Sagittaria graminea	Arrowhead	OBL	OBL	OBL	g
Sagittaria longiloba	Longbarb Arrowhead	OBL	OBL	OBL	acf
Amaranthaceae					
Amaranthus arenicola	Sandhills Pigweed	FACU	FACU–	FAC	g
Amaranthus graecizans	Prostrate Pigweed	FACW	FAC	FACU	e
Amaranthus palmeri	Palmer's Pigweed	FACU	FACU–	FACU	ad
Amaranthus retroflexus	Rough Pigweed	FACU	FACU–	FACU	af
Amaranthus rudis	Water Hemp	FACW	FAC	FACU–	g
Amaranthus spinosus	Spiny Pigweed	FACU	FACU–	——	a
Amaranthus tuberculatus	Tall Water-hemp	OBL	——	——	g
Asclepiadaceae					
Asclepias engelmanniana	Engelmann's Milkweed	——	——	——	a
Asclepias latifolia	Broadleaf Milkweed	——	——	——	a
Asclepias speciosa	Showy Milkweed	FAC	FAC	FACW–	f
Asclepias syriaca	Common Milkweed	NI	NI	——	d
Asclepias verticillata	Whorled Milkweed	——	——	——	af
Asteraceae					
Achillea millefolium	Common Yarrow	FACU	FACU	FACU	a
Ambrosia artemisiifolia	Common Ragweed	FACU	FACU–	FACU	df
Ambrosia cheiranthifolia	Ragweed	——	——	——	d
Ambrosia grayi	Bur Ragweed	FAC	FACW	——	aefg
Ambrosia psilostachya	Western Ragweed	FAC	FACU	FAC	adf
Ambrosia trifida	Giant Ragweed	FACW	FAC	FACW–	d
Antennaria parvifolia	Pussy Toes	——	——	——	a
Anthemis cotula	Dog Fennel	FACU	FACU	UPL	f
Artemisia filifolia	Sand Sage	——	——	——	a

Family/Species	Common Name	Wetland Indicator Status			Source
		Region 5	Region 6	Region 7	
Aster subulatus	Saltmarsh Aster	FACW	OBL	OBL	afg
Baccharis salicina	Willow Baccharis	FAC	FAC	FAC	cg
Berlandiera spp.	Green Eyes	——	——	——	d
Cirsium arvense	Canada Thistle	FACU	FACU	——	d
Cirsium ochrocentrum	Yellowspine Thistle	——	——	——	g
Cirsium undulatum	Wavy-leaf Thistle	FACU	FACU	FAC–	ag
Cirsium vulgare	Bull Thistle	UPL	FACU	FACU	f
Conyza canadensis	Horse-weed	FACU–	UPL	FACU	abdfg
Coreopsis tinctoria	Plains Coreopsis	FAC	FAC	FAC	acdeg
Croptilon spp.	Goldenweed	——	——	——	d
Dyssodia acerosa	Fetid Marigold	——	——	——	a
Dyssodia papposa	Fetid Marigold	——	——	——	a
Engelmannia pinnatifida	Englemann's Daisy	——	——	——	ad
Erigeron bellidastrum	Western Fleabane	——	——	——	b
Erigeron flagellaris	Fleabane	FAC	FAC	FAC–	b
Gaillardia pulchella	Indian Blanket	——	——	——	a
Grindelia squarrosa	Curly-cup Gumweed	FACU–	FACU–	FACU	adfg
Gutierrezia dracunculoides	Broomweed	——	——	——	ad
Gutierrezia sarothrae	Snakeweed	——	——	——	ab
Haplopappus ciliatus	Goldenweed	UPL	FACU	FACU	afg
Haplopappus spinulosus	Cutleaf Ironplant	——	——	——	g
Helenium amarum	Bitter Sneezeweed	FACU–	FACU	——	cf
Helenium autumnale	Sneezeweed	FACW	FACW–	FACW	c
Helenium microcephalum	Small-head Sneezeweed	NI	FACW–	FACW	a
Helianthus annuus	Common Sunflower	FACU	FAC	FAC–	afg
Helianthus ciliaris	Blueweed Sunflower	FAC	FAC	FAC	adef
Helianthus decapetalus	Sunflower	NI	NI	——	d
Helianthus petiolaris	Plains Sunflower	——	——	——	ag
Heterotheca latifolia	Camphor Weed	FACU	UPL	UPL	a
Hymenoxys odorata	Bitterweed	NI	NI	NI	ad
Iva axillaris	Poverty Weed	FAC	FAC	FAC	ab
Iva sativa	Sumpweed	——	——	——	d
Lactuca serriola	Prickly Lettuce	FAC	FAC	FAC	adg
Lygodesmia juncea	Skeletonweed	——	——	——	b
Machaeranthera bigelorvii	Hoary Aster	NI	NI	FACW	a
Machaeranthera tanacetifolia	Tahoka Daisy	——	——	——	a
Picradeniopsis woodhousii	Bahiagrass	——	——	——	acdf
Ratibida columnifera	Prairie Coneflower	——	——	——	abg
Ratibida pinnata	Grayhead Prairie Coneflower	——	——	——	g
Ratibida tagetes	Short-ray Prairie Coneflower	——	——	——	ad
Senecio douglasii	Threadleaf Groundsel	——	——	——	ad
Senecio longilobus	Groundsel	——	——	——	a

Family/Species	Common Name	Wetland Indicator Status			Source
		Region 5	Region 6	Region 7	
Senecio plattensis	Prairie Ragwort	FACU	UPL	UPL	b
Sonchus asper	Prickly Sow Thistle	FACW	FAC–	FACW	a
Sonchus oleraceus	Common Sow Thistle	FACU	UPL	UPL	d
Taraxacum officinale	Common Dandelion	FACU	FACU+	FACU	fg
Thelesperma megapotamicum	Thelesperma	——	——	——	a
Thelesperma simplifolium	Slender Greenthread	——	——	——	a
Tragopogon dubius	Goat's Beard, Salsify	——	——	——	ag
Tragopogon porrifolius	Vegetable Oyster	——	——	——	d
Vernonia marginata	Plains Ironweed	FAC	FAC	FACU	aeg
Xanthisma texanum	Sleepy Daisy	——	——	——	a
Xanthium strumarium	Cocklebur	FAC	FAC–	FAC	abcdf
Xanthocephalum spp.	Broomweed	——	——	——	d
Boraginaceae					
Amsinckia intermedia	Coast Fiddleneck	——	——	——	g
Cryptantha minima	Small Cryptantha	——	——	——	g
Heliotropium curassavicum	Seaside Heliotrope	OBL	FACW	FACW	cd
Lithospermum incisum	Narrowleaf Groomwell	——	——	——	b
Myosotis verna	Spring Forget-me-not	FAC–	FAC	FACU	g
Brassicaceae					
Camelina microcarpa	Small-Seeded False Flax	NI	NI	NI	b
Capsella bursa-pastoris	Shepherd's Purse	FACU	FAC	FAC	a
Descurainia pinnata	Tansy Mustard	——	——	——	adg
Descurainia richardsonii	Western Tansy Mustard	——	——	——	a
Dimorphocarpa palmeri	Spectacle Pod	——	——	——	a
Erysimum asperum	Western Wallflower	——	——	——	ag
Lepidium densiflorum	Peppergrass	FAC	FAC	FAC	adg
Lesquerella gordonii	Bladderpod	——	——	——	ad
Rorippa sinuata	Spreading Yellow Cress	FACW	FACW–	FACW	abcdfg
Sisymbrium altissimum	Tumbling Mustard	FACU+	FAC	FACU	ad
Thlaspi arvense	Field Pennycress	FACU	UPL	UPL	g
Cactaceae					
Echinocactus texensis	Barrel Cactus	——	——	——	a
Opuntia imbricata	Tree Cholla	——	——	——	a
Opuntia leptocaulis	Pencil Cholla	——	——	——	d
Opuntia macrorhiza	Plains Prickly Pear	——	——	——	g
Opuntia phaeacantha	Prickly Pear	——	——	——	ad

| Family/Species | Common Name | Wetland Indicator Status | | | Source |
		Region 5	Region 6	Region 7	
Caesalpiniaceae					
Cassia fasciculata	Partridge Pea	FACU	FACU–	FACU–	d
Gleditsia triacanthos	Honey Locust	FAC	FAC	FAC	d
Hoffmanseggia glauca	Indian Rush-pea	FACU	FAC	FACU	adf
Campanulaceae					
Triodanis perfoliata	Claspleaf Venus'-looking-glass	FAC	FAC–	FAC	g
Chenopodiaceae					
Allenrolfea occidentalis	Pickleweed	——	FACW	FACW	d
Chenopodium album	Lamb's Quarters	FAC	FAC	FAC–	acdefg
Chenopodium ambrosoides	Mexican Tea	FAC	FAC	FAC	b
Chenopodium berlandieri	Pitseed Goosefoot	——	——	——	g
Chenopodium glaucum	Oak-leaved Goosefoot	FACW	FAC	FAC	bf
Chenopodium incanum	Mealy Goosefoot	——	——	——	f
Chenopodium leptophyllum	Narrowleafed Goosefoot	FACU	FACU	FACU	ac
Chenopodium rubrum	Alkali Blite	FACU	——	FAC	b
Kochia scoparia	Summer Cypress	FACU	FACU	FAC	acdfg
Monolepis nuttalliana	Poverty Weed	FACW	FACU	FAC	c
Salsola iberica	Russian Thistle	FACU	FACU	FACU	acdfg
Suckleya suckleyana	Poison Suckleya	FACW	FACW	FACW	af
Clusiaceae					
Elatine triandra	Waterwort	OBL	OBL	OBL	g
Convolvulaceae					
Calystegia sepium	Hedge Bindweed	FAC	FAC–	FAC	d
Convolvulus arvensis	Field Bindweed	——	——	——	adfg
Convolvulus equitans	Bindweed	NI	NI	NI	a
Evolvulus nuttallianus	Nuttall's Bindweed	——	——	——	a
Crassulaceae					
Sedum portulaca	Stonecrop	——	——	——	f
Cucurbitaceae					
Cucurbita foetidissima	Buffalo Gourd	——	——	——	ad
Cuscutaceae					
Cuscuta gronovii	Gronovius' Dodder	——	——	——	d
Cuscuta squamata	Dodder	——	——	——	a
Cyperaceae					
Carex eleocharis	Sedge	——	——	——	g
Carex gravida	Sedge	——	——	——	g
Cyperus acuminatus	Sedge	OBL	OBL	OBL	g
Cyperus esculentus	Umbrella Sedge	FACW	FACW	FACW	adf

Family/Species	Common Name	Wetland Indicator Status			Source
		Region 5	Region 6	Region 7	
Cyperus lupulinus	Sedge	FACU	FACU	FACU	f
Cyperus oderatus	Sedge	OBL	OBL	FACW+	g
Cyperus setigerus	Sedge	FAC	FAC	NI	f
Eleocharis acicularis	Least Spikerush	OBL	OBL	OBL	cg
Eleocharis atropurpurea	Spikerush	FACW	FACW	FACW	a
Eleocharis macrostachya	Creeping Spikerush	OBL	OBL	OBL	abcdfg
Eleocharis montevidensis	Sand Spikerush	FACW	FACW+	FACW	c
Eleocharis parvula	Annual Micro-spikerush	OBL	OBL	OBL	af
Scirpus acutus	Hardstem Bulrush	OBL	OBL	OBL	acdg
Scirpus americanus	American Three-square	OBL	OBL	OBL	abd
Scirpus hallii	Hall's Bulrush	OBL	OBL	——	c
Scirpus maritimus	Alkali Bulrush	OBL	OBL	OBL	ab
Scirpus saximontanus	Bulrush	OBL	OBL	——	a
Scirpus validus	Softstem Bulrush	OBL	OBL	OBL	acdfg
Elatinaceae					
Bergia texana	Texas Bergia	OBL	OBL	OBL	cfg
Equisetaceae					
Equisetum arvense	Field Horsetail	FAC	FAC+	FACW–	d
Euphorbiaceae					
Cnidoscolus texanus	Bull Nettle	——	——	——	d
Croton capitatus	Woolly Croton	——	——	——	g
Croton dioicus	Grassland Croton	——	——	——	a
Croton spp.	Croton	——	——	——	d
Croton texensis	Texas Croton	——	——	——	b
Euphorbia albomarginata	Spurge	——	——	——	af
Euphorbia dentata	Toothed Spurge	——	——	——	afg
Euphorbia maculata	Spotted Spurge	——	——	——	g
Euphorbia marginata	Snow-on-the-mountain	FACU	FACU–	FACU–	abcdefg
Euphorbia prostrata	Prostrate Euphorbia	——	——	——	d
Euphorbia serpens	Round-leaved Spurge	——	——	——	e
Euphorbia serpyllifolia	Thyme-leaved Spurge	——	——	——	f
Stillingia sylvatica	Queen's Delight	——	——	——	d
Fabaceae					
Amorpha canescens	Leadplant	——	——	——	d
Astragalus distortus	Ozark Milk-vetch	——	——	——	d
Astragalus mollissimus	Woolly Locoweed	——	——	——	ab
Lespedeza striata	Japanese Lespedeza	UPL	FAC–	NI	d
Medicago sativa	Alfalfa	NI	NI	NI	d
Melilotus alba	White Sweet Clover	——	——	——	a
Melilotus officinalis	Yellow Sweet Clover	FACU	FACU	FACU+	adf
Oxytropis lambertii	Purple Locoweed	FACU	FACU	UPL	d

Family/Species	Common Name	Wetland Indicator Status			Source
		Region 5	Region 6	Region 7	
Oxytropis spp.	Locoweed	——	——	——	g
Psoralea lancelata	Lemon Scurf-pea	——	——	——	f
Sophora nuttalliana	White Loco	——	——	——	ab
Gentianaceae					
Eustoma grandiflorum	Prairie Gentian	FACW	FAC–	FAC	b
Geraniaceae					
Erodium cicutarium	Crane's Bill	——	——	——	a
Erodium texanum	Filaria, Stork's-bill	——	——	——	a
Haloragaceae					
Myriophyllum exalbescens	American Milfoil	OBL	NI	NI	g
Hydrophyllaceae					
Ellisia nyctelea	Waterpod	FAC	FAC	FAC	g
Iridaceae					
Sisyrinchium spp.	Blue-eyed Grass	——	——	——	a
Juncaceae					
Juncus torreyi	Torrey's Rush	FACW	FACW	FACW	f
Lamiaceae					
Salvia spp.	Sage	——	——	——	d
Teucrium canadense	American Germander	FACW	FACW–	FACW	g
Lemnaceae					
Lemna spp.	Duckweed	OBL	OBL	OBL	a
Liliaceae					
Allium drummondii	Wild Onion	——	——	——	a
Nothoscordum bivalve	False Garlic	FACU	FACU–	——	acd
Linaceae					
Linum pratense	Norton's Flax	——	——	——	a
Linum rigidum	Flax	——	——	——	g
Loasaceae					
Mentzelia nuda	Stickleaf	——	——	——	a
Lythraceae					
Ammannia auriculata	Toothcup	OBL	OBL	OBL	acg
Ammannia coccinea	Toothcup	OBL	OBL	OBL	acfg
Lythrum alatum	Winged Loosestrife	OBL	OBL	——	f
Lythrum californicum	California Loosestrife	OBL	OBL	OBL	ac

Family/Species	Common Name	Wetland Indicator Status			Source
		Region 5	Region 6	Region 7	
Malvaceae					
Hibiscus trionum	Flower-of-an-hour	——	——	——	c
Malva neglecta	Common Mallow	——	——	——	df
Malvella leprosa	Cheeseweed	FACW	FAC	FACW	acf
Malvella sagittifolia	Arrow-leaved Cheeseweed	——	——	——	ac
Sphaeralcea angustifolia	Narrowleaf Globemallow	——	——	——	a
Sphaeralcea coccinea	Scarlet Globemallow	——	——	——	ag
Sphaeralcea hastulata	Orange Globemallow	——	——	——	a
Sphaeralcea spp.	Globemallow	——	——	——	d
Marsileaceae					
Marsilea vestita	Western Water Clover	OBL	OBL	OBL	acdefg
Mimosaceae					
Desmanthus illinoensis	Illinois Bundleflower	FACU	FACU	UPL	a
Mimosa borealis	Pink Mimosa	——	——	——	ad
Mimosa strigalosa	Herbaceous Mimosa	——	FAC	——	a
Prosopis glandulosa	Honey Mesquite	UPL	FACU–	FACU	ad
Molluginaceae					
Mollugo verticillata	Carpetweed	FAC	FAC–	FAC–	e
Moraceae					
Maclura pomifera	Osage Orange	UPL	FACU	UPL	d
Morus rubra	Red Mulberry	FACU	FACU	——	d
Najadaceae					
Najas guadalupensis	Naiad	OBL	OBL	OBL	ac
Nyctaginaceae					
Mirabilis linearis	Narrowleaf Four-o'clock	NI	NI	NI	f
Onagraceae					
Gaura angustifolia	Narrowleaf Gaura	——	——	——	a
Gaura coccinea	Scarlet Gaura	——	——	——	afg
Gaura villosa	Hairy Gaura	——	——	——	a
Oenothera canescens	Spotted Evening Primrose	FACW–	FAC	FAC	abcdefg
Oenothera coronopifolia	Combleaf Evening Primrose	——	——	——	g
Orchidaceae					
Cypripedium calceolus	Yellow Lady's Slipper	FACW	FACW–	FACW	c
Osmundaceae					
Ophioglossum engelmanni	Adder's Tongue	FACU	FACU	FACW	c

Family/Species	Common Name	Wetland Indicator Status			Source
		Region 5	Region 6	Region 7	

Oxalidaceae

Oxalis dillenii	Gray-green Wood Sorrel	NI	NI	NI	d

Pedaliaceae

Proboscidea louisianica	Devil's Claw	FACU	FAC−	FAC	abdefg

Plantaginaceae

Plantago aristata	Buckhorn, Bracted Plantain	——	——	——	d
Plantago patagonica	Patagonian Plantain	UPL	FACU−	UPL	ag

Poaceae

Agropyron cristatum	Crested Wheatgrass	——	——	——	g
Agropyron smithii	Western Wheatgrass	FACU	FAC−	FAC−	abdfg
Alopecurus aequalis	Shortawn Foxtail	OBL	——	OBL	g
Alopecurus carolinianus	Carolina Foxtail	FACW	FACW	FACW	ag
Andropogon barbinodis	Cane Bluestem	——	NI	NI	a
Andropogon gerardii	Big Bluestem	FAC−	FACU	FAC−	f
Andropogon ischaemum	King Ranch Bluestem	——	——	——	a
Andropogon scoparius	Little Bluestem	FACU	FACU+	FACU	adg
Aristida oligantha	Oldfield Three-awn	——	——	——	g
Aristida pansa	Three-awn	——	——	——	a
Aristida purpurea	Red/Purple Three-awn	——	——	——	adfg
Bothriochloa ischaemum	Old World Bluestem	——	——	——	a
Bothriochloa laguroides	Silver Bluestem	——	——	——	adg
Bouteloua barbata	Sixweeks Grama	——	——	——	d
Bouteloua curtipendula	Sideoats Grama	——	——	——	abd
Bouteloua gracilis	Blue Grama	——	——	——	abdfg
Bromus japonicus	Japanese Brome	FACU	FACU	FACU	ag
Bromus secalinus	Cheatgrass	——	——	——	d
Bromus tectorum	Downy Brome	——	——	——	g
Bromus unioloides	Rescue Grass	——	——	——	a
Buchloe dactyloides	Buffalo Grass	FACU	FACU	FACU	abdefg
Cenchrus longispinus	Sandbur	——	——	——	bd
Chloris verticillata	Windmill Grass	——	——	——	adg
Cynodon dactylon	Bermuda Grass	FACU	FACU+	FACU	acd
Distichlis spicata	Saltgrass	FACW	FACW+	FACW	bf
Echinochloa colona	Jungle Rice	FACW	FACW	FACW	f
Echinochloa crusgalli	Barnyard Grass	FACW	FACW	FACW−	acdfg
Echinochloa cruspavonis	Gulf Cockspur	FACW	OBL	OBL	f
Echinochloa muricata	Rough Barnyard Grass	OBL	FACW	FACW	b
Elymus canadensis	Canadian Wild Rye	FACU	FAC+	FAC	adg
Eragrostis cilianensis	Stinkgrass	FACU	FACU	FACU+	adfg
Eragrostis curtipedicellata	Gummy Lovegrass	——	——	——	a
Eragrostis curvula	Weeping Lovegrass	——	——	——	a
Eragrostis intermedia	Plains Lovegrass	——	——	——	d
Eragrostis pectinacea	Carolina Lovegrass	FAC	FAC	FAC	a
Eragrostis pilosa	India Lovegrass	FACU	FACU	FACU	f

Family/Species	Common Name	Wetland Indicator Status			Source
		Region 5	Region 6	Region 7	
Eriochloa contracta	Prairie Cupgrass	FACU	FAC+	FACW	fg
Festuca arundinacea	Tall Fescue	FACU	FAC–	——	a
Festuca octoflora	Sixweeks Fescue	UPL	FACU	FACU	dfg
Festuca pratensis	Meadow Fescue	FAC	FACU	FACU	a
Hilaria jamessii	Galleta	——	——	——	a
Hilaria mutica	Tobosa Grass	——	——	——	d
Hordeum jubatum	Foxtail Barley	FACW	FAC+	FACW–	adg
Hordeum pusillum	Little Barley	FAC	FACU	FAC	afg
Leptochloa fascicularis	Bearded Sprangletop	OBL	FACW	FACW	adfg
Leptochloa filiformis	Red Sprangletop	——	——	——	g
Muhlenbergia arenicola	Sand Muhly	——	——	——	d
Muhlenbergia porteri	Bushy Muhly	——	——	——	a
Muhlenbergia repens	Creeping Muhly	——	——	——	a
Muhlenbergia torreyi	Ring Muhly	——	——	——	b
Munroa squarrosa	False Buffalograss	——	——	——	a
Panicum capillare	Common Witchgrass	FAC	FAC	FAC	abdg
Panicum coloratum	Kleingrass	——	——	——	a
Panicum dichotomiflorum	Fall Panicum	FAC	FACW	FAC	afg
Panicum halli	Hall's Panicum	FACU	FACU	FACU	d
Panicum miliaceum	Broom-corn Millet	——	——	——	f
Panicum obtusum	Vine Mesquite	FACW	FAC+	FAC	acdfg
Panicum virgatum	Switchgrass	FAC	FAC	FAC+	adf
Paspalum paspalodes	Knotgrass	OBL	FACW+	OBL	a
Phalaris caroliniana	May Grass	FACW	FACW	FACW	acf
Polypogon monspeliensis	Rabbitfoot Grass	OBL	FACW	FACW	a
Schedonnadrus paniculatus	Tumblegrass	——	——	——	abg
Schleropogon brevifolius	Burrograss	——	——	——	d
Setaria glauca	Yellow Foxtail	NI	FAC	FAC	g
Setaria italica	Foxtail Millet	FACU	FACU	FACU	d
Setaria verticillata	Bristly Foxtail	FAC	FAC	FACU	a
Setaria viridis	Green Foxtail	——	——	——	ad
Sitanion hystrix	Squirreltail	FACU	FACU–	UPL	ag
Sorghum halepense	Johnson-grass	FACU	FACU	FACU+	ad
Spartina pectinate	Prairie Cordgrass	FACW	FACW+	FACW	d
Sporobolus airoides	Alkali Sacaton	FAC	FAC	FAC	ad
Sporobolus asper	Rough Dropseed	FACU	UPL	UPL	dg
Sporobolus cryptandrus	Sand Dropseed	FACU–	FACU–	FACU–	adg
Sporobolus poiretii	Rattail Smutgrass	——	FACU	——	d
Tridens albescens	White Tridens	FAC	FAC–	FACU	af
Polygonaceae					
Eriogonum spp.	Wild Buckwheat	——	——	——	d
Polygonum amphibium	Water Smartweed	OBL	OBL	OBL	acdf
Polygonum argyrocoleon	Silversheath Knotweed	NI	OBL	OBL	a
Polygonum aviculare	Prostrate Knotweed	FACW	FAC+	FACW	acdefg
Polygonum hydropiperoides	Mild Waterpepper	OBL	OBL	OBL	cf

Family/Species	Common Name	Wetland Indicator Status			Source
		Region 5	Region 6	Region 7	
Polygonum lapathifolium	Pale Smartweed	OBL	FACW–	OBL	acdf
Polygonum	Pennsylvania				
pensylvanicum	Smartweed	FACW+	FACW–	OBL	acdefg
Polygonum persicaria	Lady's Thumb				
	Knotweed	OBL	FACW+	FACW+	f
Polygonum ramosissimum	Bushy Knotweed	FAC	FACW	FAC	acfg
Polygonum striatulum	Knotweed	——	FACW	——	e
Rumex altissimus	Pale Dock	FAC	FACW+	FAC+	acf
Rumex crispus	Curly Dock	FACW	FACW	FACW	abcdfg
Rumex maritimus	Golden Dock	FACW	FACW–	FACW	bg
Rumex obtusifolius	Bitter Dock	FAC	FACW–	FACW	f
Pontederiaceae					
Heteranthera limosa	Blue Mud Plantain	OBL	OBL	OBL	acefg
Heteranthera mexicana	Mud Plantain	——	OBL	——	ac
Heteranthera peduncularis	Mud Plantain	——	——	——	g
Portulacaceae					
Portulaca mundula	Shaggy Portulaca	NI	NI	NI	a
Portulaca oleracea	Common Purslane	FAC	FAC	FAC	aefg
Portulaca suffrutescens	Purslane	——	——	——	f
Talinum parviflorum	Prairie Fameflower	——	——	——	g
Potamogetonaceae					
Potamogeton foliosus	Leafy Pondweed	OBL	OBL	OBL	c
Potamogeton natans	Floatingleaf				
	Pondweed	OBL	OBL	OBL	cd
Potamogeton nodosus	Longleaf Pondweed	OBL	OBL	OBL	af
Potamogeton pectinatus	Sago Pondweed	OBL	OBL	OBL	acd
Ranunculaceae					
Myosurus minimus	Mousetail	FACW	FACW+	OBL	cefg
Ranunculus cymbalaria	Shore Buttercup	OBL	OBL	OBL	c
Ranunculus sceleratus	Cursed Crowfoot	OBL	OBL	OBL	cf
Rosaceae					
Potentilla rivalis	Brook Cinquefoil	FACW+	FACW+	OBL	ac
Rubiaceae					
Diodia teres	Rough Buttonweed	FACU	FACU–	FACU–	d
Ruppiaceae					
Ruppia maritima	Widgeon Grass	OBL	OBL	OBL	a
Salicaceae					
Populus deltoides	Cottonwood	FAC	FAC	FACW–	ad
Salix amygdaloides	Peach-leaf Willow	FACW	FACW	FACW	af
Salix exigua	Sandbar Willow	OBL	FACW+	OBL	a
Salix nigra	Black Willow	OBL	FACW+	——	acf

Family/Species	Common Name	Wetland Indicator Status			Source
		Region 5	Region 6	Region 7	
Scrophulariaceae					
Bacopa rotundifolia	Water Hyssop	OBL	OBL	OBL	acfg
Castilleja integra	Indian Paintbrush	——	——	——	b
Linaria vulgaris	Butter-and-eggs	——	——	——	d
Veronica peregrina	Purslane Speedwell	OBL	OBL	OBL	cefg
Solanaceae					
Physalis heterophylla	Clammy Ground Cherry	——	——	——	d
Physalis viscosa	Yellow Ground Cherry	——	——	——	a
Quincula lobata	Purple Ground Cherry	NI	NI	NI	ag
Solanum elaeagnifolium	Silver-leaf Nightshade	——	——	——	adg
Solanum interius	Plains Black Nightshade	——	——	——	f
Solanum rostratum	Buffalo Bur	——	——	——	abdefg
Tamaricaceae					
Tamarix chinensis	Chinese Tamarix	FACW	FACW	FACW	b
Tamarix gallica	French Tamarix	——	FACW–	NI	ad
Typhaceae					
Typha angustifolia	Narrow-leaved Cattail	OBL	OBL	OBL	bfg
Typha domingensis	Southern Cattail	OBL	OBL	OBL	acf
Typha latifolia	Broad-leaved Cattail	OBL	OBL	OBL	ad
Ulmaceae					
Celtis occidentalis	Hackberry	FACU	FAC	NI	d
Ulmus pumila	Siberian (Chinese) Elm	UPL	NI	NI	ad
Urticaceae					
Urtica gracilis	Slim Nettle	FACW	FAC	FACW	d
Verbenaceae					
Lippia cuneifolia	Wedge-leaf Frog-fruit	FAC	FAC	FACW	abcfg
Lippia lanceolata	Northern Frog-fruit	OBL	FACW	OBL	c
Lippia nodiflora	Frog-fruit	FACW	FAC	FACW	ade
Verbena bracteata	Prostrate Vervain	FACU	FAC	FAC	abcdefg
Violaceae					
Viola rafinesquii	Johnny-jump-up	——	——	——	g
Zannichelliaceae					
Zannichellia palustris	Horned Pondweed	OBL	OBL	OBL	c
Zygophyllaceae					
Tribulus terrestris	Goathead	——	——	——	adfg

LITERATURE CITED

Allen, B.L., B.L. Harris, K.R. Davis, and G.B. Miller. 1972. The mineralogy and chemistry of High Plains playa lake soils and sediments. Water Resources Center Report WRC-72-4, Texas Tech University, Lubbock. 75 pp.

Anderson, A. M. 1997. Habitat use and diet of amphibians breeding in playa wetlands on the Southern High Plains of Texas. M.S. Thesis, Texas Tech University, Lubbock. 119 pp.

Berthelsen, P.S., L.M. Smith, and C.L. Coffman. 1989. CRP land and game bird production in the Texas High Plains. Journal of Soil and Water Conservation 44:504-507.

Bolen, E.G., L.M. Smith, and H.L. Schramm, Jr. 1989. Playa lakes: prairie wetlands of the Southern High Plains. BioScience 39:615-623.

Curtis, D. and H. Beierman. 1980. Playa lakes characterization study. U.S. Fish and Wildlife Service, Fort Worth, Texas. 56 pp.

Cushing, C.E., R.R. Mazaika, and R.C. Phillips. 1993. Ecological investigations at the Pantex plant site. Battelle, Pacific Northwest Laboratory, U.S. Department of Energy Contract DE-AC06-76RLO 1830, Amarillo, Texas. 28 pp.

Flickinger, E. L. and A. J. Krynitsky. 1987. Organochlorine residues in ducks on playa lakes of the Texas Panhandle and eastern New Mexico. Journal of Wildlife Diseases 23:165-168.

Great Plains Flora Association. 1991. Flora of the Great Plains. University Press of Kansas, Lawrence. 1402 pp.

Grubb, H.W. and D.L. Parks. 1968. Multipurpose benefits and costs of modifying playa lakes of the Texas High Plains. International Center for Arid and Semi-Arid Land Studies, Special Report No. 6, Texas Tech University, Lubbock. 66 pp.

Gustavson, T.C., V.T. Holliday, and S.D. Hovorka. 1995. Origin and development of playa basins, sources of recharge to the Ogallala aquifer, Southern High Plains, Texas and New Mexico. Report of Investigations No. 229, Bureau of Economic Geology, University of Texas, Austin. 44 pp.

Guthery, F. S. and F. C. Bryant. 1982. Status of playas in the Southern Great Plains. Wildlife Society Bulletin 10:309-317.

Guthery, F.S., J. Custer, and M. Owen. 1980. Texas panhandle pheasants: their history, habitat needs, habitat development opportunities, and future. U.S. Department of Agriculture, Forest Service, Rocky Mountain Forest and Range Experiment Station, General Technical Report, RM-74, Fort Collins, Colorado. 11 pp.

Guthery, F. S., F. C. Bryant, B. Kramer, A. Stoecker, and M. Dvorack. 1981. Playa assessment study. U.S. Water and Power Resources Service, Southwest Region, Amarillo, Texas. 118 pp.

Guthery, F.S., J.M. Pates, and F.A. Stormer. 1982. Characterization of playas of the north-central Llano Estacado in Texas. Transactions of the North American Wildlife and Natural Resources Conference 47:516-527.

Haukos, D.A. and L.M. Smith. 1993. Seed bank composition and predictive ability of field vegetation in playa lakes. Wetlands 13:32-40.

Haukos, D.A. and L.M. Smith. 1994a. The importance of playa wetlands to biodiversity of Southern High Plains. Landscape and Urban Planning 28:83-98.

Haukos, D.A. and L.M. Smith. 1994b. Composition of seed banks along an elevational gradient in playa wetlands. Wetlands 14:301-307.

Haukos, D.A. and L.M. Smith. 1996. Effects of moist-soil management on playa wetland soils. Wetlands 16:143-149.

Hertal, L.D. and R.K. Smith. 1994. Urban playa lake management. Pages 109-112 in L.V. Urban and A.W. Wyatt, eds. Proceedings of the playa basin symposium. Water Resources Center, Texas Tech University, Lubbock.

Hoagland, B. 1991. Final report, Colorado playa lake study. The Nature Conservancy, Denver, Colorado. 21 pp.

Holliday, V.T. 1991. The geologic record of wind erosion, eolian deposition, and aridity on the Southern High Plains. Great Plains Research 1:7-25.

Johnston, M.C. 1995. Floristic survey of the Pantex plant site, Carson County, Texas. U.S. Department of Energy, Amarillo, Texas. 45 pp.

Kartesz, J.T. 1994. A synonymized checklist of the vascular flora of the United States, Canada, and Greenland, 2nd edition, Vol. 1 — checklist. Timber Press, Portland, Oregon. 622 pp.

Kindscher, K. 1994. Vegetation of western Kansas playa lakes —1994. U.S. Department of Agriculture, Natural Resources Conservation Service, Salina, Kansas. 9 pp.

Kindscher, K. and C. Lauver. 1993. Preliminary vegetation analysis of western Kansas playa lakes. U.S. Department of Agriculture, Natural Resources Conservation Service, Salina, Kansas. 6 pp.

Luo, H.R., L.M. Smith, B.L. Allen, and D.A. Haukos. 1997. Effects of sedimentation on playa wetland volume. Ecological Applications 7:247-252.

Mollhagen, T.R., L.V. Urban, R.H. Ramsey, A.W. Wyatt, C.D. McReynolds, and J.T. Ray. 1993. Assessment of non-point source contamination of playa basins in the High Plains of Texas. Water Resources Center, Texas Tech University, Lubbock. 23 pp.

National Wetland Inventory. 1996. "National List of Plant Species that Occur in Wetlands." U.S. Fish and Wildlife Service, St. Petersburg, Florida (database).

Nelson, R.W., W.J. Logan, and E.C. Weller. 1983. Playa wetlands and wildlife on the Southern Great Plains: a characterization of habitat. FWS/OBS 83/28, U.S. Fish and Wildlife Service, Washington, D.C. 163 pp.

Osterkamp, W.R. and W.W. Wood. 1987. Playa-lake basins on the Southern High Plains of Texas and New Mexico: part I. hydrologic, geomorphic, and geologic evidence for their development. Geological Society of America Bulletin 99:215-223.

Price, D.J., B.R. Murphy, and L.M. Smith. 1989. Effects of tebuthiuron on characteristic green algae found in playa lakes. Journal of Environmental Quality 18:62-66.

Reddell, D.L. 1965. Water resources of playa lakes. Cross Section 12:1.

Reed, E.L. 1930. Vegetation of the playa lakes in the Staked Plains of western Texas. Ecology 11:597-600.

Rowell, C.M., Jr. 1971. Vascular plants of the playa lakes of the Texas Panhandle and South Plains. Southwestern Naturalist 15:407-417.

Rowell, C.M., Jr. 1981. The flora of playas. Pages 21-29 in J.S. Barclay and W.V. White, eds. Playa Lakes Symposium. U.S. Fish and Wildlife Service, FWS/OBS-81/07, Fort Worth, Texas.

U.S. Fish and Wildlife Service. 1988. Playa lakes region waterfowl habitat concept plan, category 24 of the North American Waterfowl Management Plan. U.S. Fish and Wildlife Service, Albuquerque, New Mexico. 37 pp.

Wallace, B. M. 1984. Organochloride pesticide residues in pheasant eggs and waterfowl from the Texas High Plains. M.S. Thesis. Texas Tech University, Lubbock. 30 pp.

Wood, W.W. and W.R. Osterkamp. 1987. Playa-lake basins on the Southern High Plains of Texas and New Mexico: part II. a hydrologic and mass-balance arguments for their development. Geological Society of America Bulletin 99:224-230.

Zartman, R.E. 1987. Playa lakes recharge aquifers. Crops Soils 39:20.

Zartman, R.E., R.H. Ramsey, P.W. Evans, G. Koenig, C. Truby, and L. Kamara. 1996. Outer basin, annulus, and playa basin infiltration studies. Texas Journal of Agriculture and Natural Resources 9:23-32.

GLOSSARY

Abaxial: Located on the side of an organ away from the axis.

Achene: A dry, indehiscent, 1-seeded fruit.

Acuminate: Gradually tapering to sharp, terminal point.

Acute: Terminating in a sharp point.

Adventitious: Developing irregularly or accidentally.

Alternate: Located singly one above the other on the axis.

Annual: Completing life cycle within 1 year or 1 season.

Apex: The tip of a structure.

Appressed: Flatly and closely pressed against.

Aromatic: Having a fragrant odor.

Articulate: Having nodes or joints.

Ascending: Sloping or rising upward.

Attenuate: Gradually tapering to very slender tip.

Auricle: An ear-shaped appendage at the junction of the blade and sheath in some grasses.

Auriculate: Having auricles.

Awn: A terminal, bristlelike appendage at the end of an organ.

Axil: An angle formed between two organs (e.g., petiole and stem).

Axis: The main longitudinal support on which parts are arranged.

Beak: A long, firm, slender point.

Berry: A pulpy, indehiscent fruit with few to many seeds.

Biennial: Living for 2 years.

Bifid: Two-cleft.

Bilateral: Two-sided; structures on 2 sides of an organ.

Bipinnate: Twice compound, with leaflets arranged on both sides of the axis.

Bivalved: Having two valves.

Blade: The expanded portion of a leaf.

Bristle: A stiff hair.

Bur: A rough or prickly covering surrounding the fruits or spikelets of some genera.

Bract: A modified, reduced leaf.

Caducous: Falling off early.

Callous: Having the texture of a callus, which is a hard protuberance.

Calyx: The outer whorl of the perianth, composed of sepals.

Canescent: Having gray or whitish pubescence.

Capsule: A dry, dehiscent fruit with > 1 carpel.

Catkin: A spikelike inflorescence of unisexual, apetalous, bracteate flowers.

Caudex: The tough, persistent base of an otherwise herbaceous stem; the main axis of a plant consisting of root and stem.

Caulescent: Having an obvious stem.

Cauline: Belonging to the stem.

Chaffy: A condition of a head inflorescence in the Asteraceae in which individual flowers are subtended by a thin, dry bract.

Ciliate: Fringed with marginal hairs.

Claviform: Club-shaped, or thickened toward the top.

Clone: A group of individuals of the same genotype, usually propagated vegetatively.

Columnar: Column-shaped.

Compound: Made up of ≥ 2 parts.

Compressed: Flattened strongly, typically laterally; keeled.

Connate: Joined or united, usually similar structures.

Conical: Cone-shaped.

Contracted: Narrow or dense, with short or appressed branches.

Cordate: Heart-shaped, with rounded lobes and a sinus at the base.

Corm: The fleshy, bulblike base of a stem, usually underground.

Corolla: The inner whorl of the perianth, composed of the petals.

Crisped: Curled.

Crown: The portion of a stem at the surface of the soil.

Culm: The stem of grass or sedge.

Cyathia: An inflorescence with a cuplike involucre bearing unisexual flowers (Euphorbia).

Cylindrical: Shaped like a tube, round in cross-section with parallel margins.

Cyme: A broad, flat inflorescence with central flower blooming first.

Decumbent: Reclining, but with the terminus ascending.

Dehiscent: Opening regularly by slits, valves, etc.

Deltoid: Triangular.

Dentate: With sharp, spreading teeth.

Denticulate: Minutely dentate.

Diffuse: Widely spreading.

Dimorphic: Occurring in 2 forms.

Dioecious: Having staminate and pistillate flowers on different plants.

Disarticulate: Separated at the joints naturally at maturity.

Disk: A fleshy or elevated development of the receptacle; in Asteraceae the central portion of the flowering head.

Dorsal: Referring to the back or outer surface of an organ; the lower surface of a leaf.

Ellipsoid: Oval in shape, widest at middle and tapering equally to both rounded ends.

Elongate: Narrow, the length several times the width or thickness.

Entire: Whole; with a continuous margin.

Erect: Upright in relation to the ground; perpendicular to the ground.

Farinose: Covered with meal-like powder.

Filiform: Long and very slender; threadlike.

Floret: Small flower of dense inflorescence.

Fruit: A ripened ovary, along with any other structures ripening and forming a unit with it.

Gamopetalous: Having the petals at least partially united; sympetalous.

Geniculate: Bent sharply, like a knee.

Glabrate/Glabrescent: Nearly smooth, or becoming smooth, not hairy.

Glabrous: Smooth and not hairy.

Glandular: Having glands or secretory organs.

Glaucous: Covered with a whitish waxy bloom that rubs off easily.

Globose: Round or spherical.

Globular: Spherical in shape.

Glomerule: A small, compact cluster.

Glume: A small, chaffy bract; a sterile bract at the base of a grass spikelet.

Granulate: Covered with minute, grainlike particles.

Head: A short, dense cluster of sessile or nearly sessile flowers.

Hemispherical: Half-circular.

Herbaceous: Having the character of an herb; a plant lacking persistent woody parts aboveground.

Hirsute: Having coarse or stiff, long hairs.

Hispid: Having bristly or rigid hairs.

Imbricate: Overlapping, as do shingles on a roof.

Imperfect: Having either functional stamens or functional pistils, not both; unisexual.

Incised: Cut irregularly, sharply, and more or less deeply.

Indehiscent: Not opening by sutures, pores, etc.

Indurate: Hard.

Inflorescence: The mode of arrangement of flowers, or the flowering portion of the plant.

Involucre: A whorl of bracts under a flower or flower cluster.

Involute: Rolled inward, toward the upper side.

Keel: A central dorsal ridge; the 2 united front petals of a flower (Fabaceae).

Lanceolate: Lance-shaped; much longer than broad, widest near the base and tapering to the apex.

Latex: A milky sap.

Leaflet: One part (blade) of a compound leaf.

Legume: A bilaterally symmetrical fruit produced from a unilocular ovary, dehiscing into 2 valves, with seeds attached along with ventral suture (Fabaceae).

Lemma: The lower of 2 bracts enclosing the flower of a grass.

Lenticular: Lens-shaped.

Ligule: A strap-shaped limb or body.

Lobe: A partial division of an organ, especially if rounded.

Median: Pertaining to the middle.

Membranaceous: Thin, pliable, more or less translucent.

Midrib: Central vein of a leaf or leaflet.

Monecious: Having staminate and pistillate flowers on the same plant.

Nerve: A simple vein or rib.

Node: The place on the stem where a leaf is borne.

Nutlet: A small nut.

Oblique: Slanting.

Oblanceolate: Lanceolate with broadest part above the middle and tapering toward the base.

Obovate/Obovoid: Egg-shaped, with broader part toward the top.

Opposite: Arranged 2 at each node, on opposite sides of the axis.

Orbicular: Circular.

Ovate-oblong: Egg-shaped with the broader part near the base.

Ovoid: Oval in flat outline.

Palmate: Divided in a palmlike or handlike manner.

Panicle: An irregularly compound inflorescence with pedicellate flowers.

Paniculate: Borne in a panicle.

Papillose: Having minute, nipple-shaped projections on a surface.

Pappus: Modified perianth forming a crown on an achene (Asteraceae).

Pedicel: The stalk of a single flower.

Pedicellate: Borne on a pedicel.

Peduncle: The stalk of a flower cluster or of 1 flower when it is the only member of an inflorescence.

Pendulous: Hanging down.

Perennial: Growing > 2 years; completing several reproductive cycles.

Perfect: Having both functional stamens and pistils.

Perianth: The floral envelope, corolla and calyx.

Pericarp: The fruit wall.

Petal: A modified floral leaf, usually colored other than green, located between the calyx and the stamens.

Petiolate: With a petiole.

Petiole: A stalk or stem of a leaf.

Pilose: Having long, soft, spreading hairs.

Pinnae: A primary leaflet or division of a pinnately compound leaf.

Pinnate: Having 2 rows of lateral divisions along the main axis.

Pinnatifid: Deeply cut in a pinnate fashion but not entirely to the main axis.

Pinnatisect: Pinnately divided to the midrib.

Pistillate: Possessing ≥ 1 pistils but lacking stamens.

Plumose: Feathery, with a long pubescence or with pinnately arranged bristles.

Pod: A general term applied to any type of fruit that dehisces.

Prickle: A small, sharp outgrowth of epidermis.

Prostrate: Lying flat on the ground.

Puberulent: Minutely pubescent with hairs hardly visible.

Pubescent: Covered with short, soft, downy hairs; a general term for any kind of hairiness.

Punctate: Having colored or trans-lucent dots or pits.

Raceme: A flower cluster that is elongate with short, lateral branches (pedicels) on which the individual flowers are borne. The older flowers are below, the younger ones above.

Racemiform: Having the appearance of a raceme but not necessarily the specific structure.

Racemose: An inflorescence that appears like a raceme and develops in an indeterminate way, the older flowers below, the younger flowers above.

Rachis: The axis of a spike, spicate raceme, or raceme inflores-cence; the axis of a compound leaf.

Ray: Outer floret of Asteraceae, with straplike corolla, no stamens, functionally pistillate; the branch of an umbel.

Receptacle: The expanded end of the axis bearing flower parts.

Recurved: Curved backward or downward.

Reniform: Kidney-shaped.

Resinous: Producing or containing resin.

Rhizomatous: Possessing a rhizome.

Rib: One of the main longitudinal veins on a leaf or other structure.

Rotate: Wheel-shaped.

Rosette: A cluster of organs arranged in a compact circle.

Saccate: Bag-shaped.

Saggitate: Shaped like an arrowhead, with basal lobes pointing downward.

Scapose: Bearing flowers on a scape.

Scabrous: Rough to the touch; caused by short, stiff, angled hairs on the surface.

Scale: Any thin, dry, appressed organ (usually leaf or bract).

Scurfy: With scalelike particles.

Sepal: One division of the calyx.

Serrate: With sharp teeth pointing forward.

Serrulate: Finely serrate.

Sessile: Without a stalk.

Setae: Bristles.

Sheath: A tubular structure surrounding part or all of an organ; the portion of a grass leaf that surrounds the stem.

Silique: An elongate, dry, dehiscent fruit with a septum separating the 2 valves (Brassicaceae).

Sinuate: With margin strongly wavy.

Solitary: Occurring singly; alone.

Sori: A cluster of sporangia on a fern frond.

Spathe: A leaflike bract surrounding an inflorescence (Araceae) or smaller bract below flower (Iris).

Spatulate: Spoon-shaped.

Spicate: Spikelike.

Spiciform: Shaped like a spike.

Spike: A simple, elongated inflorescence with sessile flowers.

Spikelet: A small or secondary spike.

Spine: A sharp, rigid, outgrowth, usually from the wood of a stem.

Spinose: Having spines.

Sporocarp: An organ containing sporangia.

Stamen: The pollen-bearing organ of a flowering plant.

Staminate: Having stamens but no functional pistil.

Stellate: Star-shaped.

Stigma: The part of the pistil that receives the pollen.

Stolon: A horizontal stem that roots at the tip or at the nodes; runner.

Striate: Marked with fine longitudinal lines or ridges.

Strigulose: Having minute, sharp, stiff, straight and appressed hairs that are often swollen at base.

Style: The usually elongated part of the pistil between the ovary and the stigma.

Sub: A prefix meaning "beneath," but sometimes signifying "slightly" or "somewhat."

Succulent: Fleshy, juicy.

Taproot: The primary descending root.

Terete: Circular in transverse section.

Terminal: Borne at or belonging to the extremity or summit; distal.

Throat: The opening of a gamopetalous (petals at least partially united) corolla.

Toothed: Having small, pointed marginal projections.

Trichomes: A hair or bristle growing from the epidermis.

Trigonous: Three-angled, with plane faces between.

Trifoliate: Having 3 leaflets.

Trullate: Trowel-shaped, with straight margins and widest below the middle.

Truncate: Ending abruptly, as if cut off nearly straight across.

Tuber: A thick, fleshy to hard subterranean organ with numerous buds.

Tufted: Clustered.

Umbel: A flat-topped or rounded inflorescence in which pedicels or peduncles arise from a common point (Apiaceae).

Unilateral: One-sided.

Valve: A separable part; one of the units into which a capsule splits.

Vein: A thread of fibrovascular tissue in a leaf.

Verticel: A whorl.

Vesicle: A small cavity or bladder.

Villous: Having long, soft hairs, not matted.

Virgate: Long, slender, and straight; wand-shaped.

Whorled: Arranged in a circle, as leaves around the stem at a single node.

Wing: A thin, membranaceous extension of an organ; the lateral petal of a Fabaceae flower.

Woolly: Having curly, soft hairs, usually matted; lanate.

Zygomorphic: Irregular; divisible into equal halves in only 1 plane.

MEASUREMENT CONVERSION TABLE

EQUIVALENTS FOR CONVERSIONS FROM TEXT

Metric Unit	=	U.S. System Equivalent
Meter (m)	=	39.37 inches or 3.28 feet
Decimeter (dm)	=	3.937 inches
Centimeter (cm)	=	0.3937 inch
Millimeter (mm)	=	0.03937 inch
Kilometer (km)	=	0.62137 miles
Hectare (ha)	=	2.471 acres

INDEX TO COMMON AND SCIENTIFIC NAMES